New Uses of Sulfur—II

New Uses of Sulfur—II

Douglas J. Bourne, EDITOR

Duval Corporation

A symposium sponsored by

the Division of Industrial

and Engineering Chemistry

at the 173rd Meeting of the

American Chemical Society,

New Orleans, La.,

March 22–23, 1977

ADVANCES IN CHEMISTRY SERIES **165**

AMERICAN CHEMICAL SOCIETY

WASHINGTON, D. C.   1978

Library of Congress CIP Data

Symposium on Sulfur Utilization: a Progress Report,
New Orleans, La., 1977.
New uses of sulfur, II.
(Advances in chemistry series; 165 ISSN 0065-2393)

Bibliography: p.
Includes index.

1. Sulphur—Congresses.
I. Bourne, Douglas J. II. American Chemical Society.
Division of Industrial and Engineering Chemistry. III.
Title. IV. Series: Advances in chemistry series; 165.

QD1.A355 no. 165 [TP245.S9] 540'.8s [661.'07'23]
ISBN 0-8412-0391-1   78-1004   ADCSAJ 165
                                     1–283 1978

Advances in Chemistry Series

Robert F. Gould, *Editor*

FOREWORD

ADVANCES IN CHEMISTRY SERIES was founded in 1949 by the American Chemical Society as an outlet for symposia and collections of data in special areas of topical interest that could not be accommodated in the Society's journals. It provides a medium for symposia that would otherwise be fragmented, their papers distributed among several journals or not published at all. Papers are reviewed critically according to ACS editorial standards and receive the careful attention and processing characteristic of ACS publications. Volumes in the ADVANCES IN CHEMISTRY SERIES maintain the integrity of the symposia on which they are based; however, verbatim reproductions of previously published papers are not accepted. Papers may include reports of research as well as reviews since symposia may embrace both types of presentation.

CONTENTS

PREFACE

For many years scientists and engineers have attempted to modify the properties of sulfur so that it might be of value as a construction material, but until recently these random efforts lacked direction. Frequently the objectives of work described in many published papers could not be reconciled with the experimental design. For example, authors cited availability, low cost, and low toxicity as reasons for evaluating sulfur as a construction material. The same authors would then proceed to modify sulfur with additives which were in short supply, expensive, and highly toxic. The aging characteristics of sulfur concretes beyond a 28-day period were never examined, and durability outside the laboratory environment was seldom considered. With these limitations, the data were of little value to the construction industry.

In 1974, the American Chemical Society and The Sulphur Institute organized a symposium entitled, "Sulfur Utilization—A Status Report." It brought together men of different disciplines to permit them to examine simultaneously the properties of sulfur and the needs of the construction industry. These presentations and discussions encouraged the different groups to initiate or enlarge their efforts in this area.

In the spring of 1977, the "Symposium on Sulfur Utilization: A Progress Report" was organized as a follow-up to report the recent accomplishments which take us closer to the utilization of sulfur in the construction of chemical and mineral processing equipment, roads, houses, etc. Although a two-day symposium cannot cover fully all the important work carried out during the three-year interim, I am convinced that the informal and frequently lively exchanges which ensued will go far toward hastening the use of this technology. The text of this volume is based upon the papers presented at the Symposium.

I wish to express my thanks to The Sulphur Institute staff and others who were vital to the organization of the meeting.

Duval Corporation   Douglas J. Bourne
Houston, TX 77001
July 1977

The Scientific Basis for Practical Applications of Elemental Sulfur

MAX SCHMIDT

Institut für Anorganische Chemie der Universität Würzburg, Am Hubland, D-8700 Würzburg, West Germany

Because of its unique chemical and physical properties, elemental sulfur has interesting potential applications, particularly in the construction industry. Despite years of research and developmental activities, little sulfur has been consumed in these applications. One reason for this is that sulfur atoms combine with each other to form the extremely complicated and complex system of chain or ring molecules, S_x. Depending on x, the physical and chemical properties of sulfur molecules and the molecular equilibria mixtures change rather drastically. For practical applications, the correlations between molecular size, molecular geometry, chemical and physical stability, and other common properties of sulfur must be known. Stereochemical aspects may indicate the performance of S_x in different applications.

We can follow the history of sulfur back to the days of Sodom and Gomorrah. Nowadays, sulfur is a cheap material which is available in enormous quantities industrially in a purity that is exceptionally high for an element. Only a negligibly small proportion of this production is used directly as the element mainly because we still know and understand far too little about the properties of sulfur to be able to change them to meet selected practical ends. For example, scientists still dipute the melting point of sulfui.

What actually is sulfur? Is that a real question? Certainly not if we are content with the answer, trivial today, that it has the simplest stoichiometric composition conceivable—merely a series of atoms of mass number 16. This answer, however, clarifies only the centuries-long arguments about the composition of sulfur—whether this was a combination of one

0-8412-0391-1/78/33-165-001$05.00/0

"fiery" principle with one "acid" principle—a matter that was connected very intimately with the development of chemistry from ancient times through alchemy to the present day. Nevertheless, this answer is of no further help if we want to know more about the properties of sulfur. Actually, this element presents an extremely complicated chemical system which is still far from understood. According to the external conditions, that system is comprised of sulfur molecules, each containing from two to several hundred thousand atoms mixed together in great variety: $S_x(x = 2$ to $\approx 10^6 \, !)$.

Reactivity of Sulfur Atoms

We are dealing here with true dynamic equilibrium mixtures that change their molecular composition on dissolution and by forming new chemical bonds. Sulfur atoms have a surprisingly well developed tendency and ability to combine with one another to form chain or ring molecules. In this respect only carbon is superior to sulfur. The only two homonuclear singular bond energies higher than that for the sulfur–sulfur bond (265 kJ/mol) are those for hydrogen (435 kJ/mol) and for carbon with (330 kJ/mol) (1).

Sulfur atoms have six outer electrons, with the two unpaired $3p$ electrons covalently bonded to neighboring atoms (Figure 1). Therefore

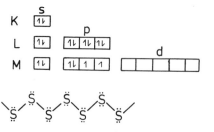

Figure 1. (above) Electronic configuration of a sulfur atom in the ground state; (below) two-dimensional representation of a three-dimensional sulfur helix

the resulting chains cannot be collinear but must be zig-zag. Such zig-zag chains, of course, are not planar, because each sulfur atom still has two further electron pairs which prevent free rotation around the sulfur–sulfur bond, producing a favored dihedral angle (Figure 2).

According to whether we proceed upwards or downwards when passing from the third to the fourth atom in the middle plane, we have the beginning of a right- or a left-handed helix. Sulfur chains, and, of course also all the many sulfanes and sulfane-derivatives, i.e., compounds containing chains of sulfur atoms, thus must have helical conformations. A helix with zero translation leads to a non-planar ring.

Thus, the main factors determining the conformation of compounds with isolated or cumulated sulfur–sulfur bonds are the bond distance d,

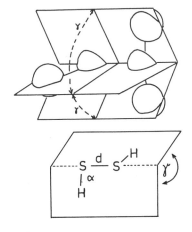

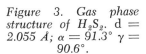

Figure 2. *Dihedral angle (γ) between any three neighboring four sulfur atoms which begin a right- or a left-handed helix*

Figure 3. *Gas phase structure of H_2S_2. d = 2.055 Å; α = 91.3° γ = 90.6°.*

bond angle α, and dihedral angle γ. This is illustrated by the simplest molecule with an isolated sulfur–sulfur bond, disulfane H_2S_2 (Figure 3).

Geometrical Aspects of Sulfur–Sulfur Bonds

In the following discussion, the different arguments regarding the theory of the sulfur–sulfur bond, such as participation of d-orbitals, double bond character, force constants, barrier of rotation, etc. (*1*) are not covered.

In disulfane a rotation barrier of between 8 and 29 kJ/mol has been calculated by different authors. Because there is practically no steric hindrance between the two hydrogen atoms, in this simplest molecule the sulfur–sulfur bond distance is 2055 Å—a nearly ideal bond—and the dihedral angles are 91.3° and 90.6°, respectively (*1*).

If we pass from the simple case of isolated to cumulated sulfur–sulfur bonds, then interactions between next nearest neighbors will influence the exact geometry of the molecules significantly (Figure 4). The simplest chemical compound containing cumulated sulfur–sulfur is elemental sulfur itself.

In spite of the complexity of the system, elemental sulfur and its incurable, confused nomenclature—literally almost the whole Greek alphabet has been misused to denote mixtures of largely unknown composition—there is one very important fact in this system. It is the fact that

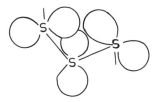

Figure 4. *Interaction of 3pπ orbitals with nearest neighbors in sulfur chains*

under normal conditions only one compound of sulfur atoms is thermodynamically stable (Figure 5), namely the eight-membered cyclooctasulfur, S_8. For reasons mentioned before, this ring naturally is not planar.
The crown-shaped molecules crystallize in an orthorhombic structure up

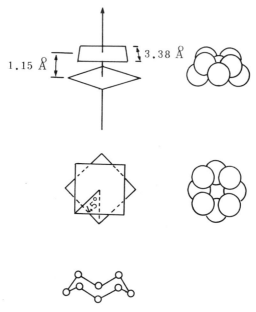

Figure 5. Structure of cyclooctasulfur. Sulfur–sulfur distance ≈ 2.06 Å; bond angle
β ≈ 108°; dihedral angle γ = 98°.

to 95.4°C where they pass into monoclinic crystals. These eight-membered rings decompose partially at ~ 119°C, forming other molecules
which depress the solidification temperature; sulfur has no melting point
in the correct sense of the word. The sulfur–sulfur distance in this ring is
2.06 Å, the bond angle is 108°, and the dihedral angle is 98°. The difference of these latter two angles from the ideal of 90°, as in disulfane
with its isolated sulfur–sulfur bond in the system with eight cumulated
bonds, is caused mainly by interactions between next nearest neighbors.
If there are at least two cumulated sulfur–sulfur bonds with two other
ligands R (Figure 6) or two more sulfur atoms (at least five sulfur atoms
altogether), then there are three different possible conformations, namely
cis, d-trans, and l-trans. On storing, all the other sulfur molecules are
converted into this most stable form at room temperature.

Cyclohexasulfur, S_6, also has an all-cis arrangement (Figure 7). It
was the only other definitely known sulfur molecule besides S_8 at the
beginning of our synthetic work. It was considered as a sort of outsider

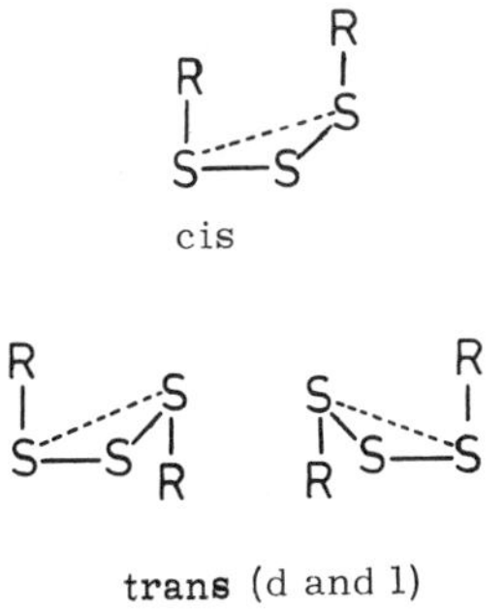

R
R
|
S—S

cis

R   R
| S   S |
S—S | S—S
     R   R

trans (d and l)

Figure 6. Possible conformations with respect to cumulated sulfur–sulfur bonds

in sulfur chemistry for ~ 80 years and was known to be one of the decomposition products of thiosulfuric acid. The properties of S_6 convincingly demonstrate the fact that in symmetrical ring molecules of sulfur atoms, bond angles and dihedral angles are not independent but are interconnected. At a given bond angle α, ring closure is possible only at a certain dihedral angle γ; the latter one is determined by the ring size and the number of cis, d-trans, and/or l-trans arrangements. Therefore sometimes the dihedral angle is not optimum, for example, at the minimum of the torsion potential, as with S_6. As in S_8, the bond distance is ~ 2.06 Å. The bond angle is 101°, somewhat smaller than that for S_8 (108°), but γ is only 75° while for the stable cyclooctasulfur, it is 98°. This results in a rather high ring strain in the chair formed six-membered ring. The very light-sensitive cyclohexasulfur is decomposed by visible light as well as by moderate warming up to ~50°C. Chemically it reacts with both nucleophiles and electrophiles in redox reactions that are up to 10,000 times faster than those of S_8. Since the nature of this sulfur allotrope, previously known as "Engel's sulfur" or "Aten's sulfur," was recognized, there have been many attempts to prepare sulfur rings other than S_8 and S_6; prior to our work, they were all unsuccessful.

Stereochemical Aspects of Even-Numbered Sulfur Rings

In view of this rather unsatisfactory situation we asked ourselves whether it might be possible to build new thermodynamically unstable

Figure 7. Structure of cyclohexasulfur. Sulfur–sulfur distance ≈ 2.06 Å; bond angle β ≈ 102°; dihedral angle γ ≈ 75°.

molecules containing atoms of only one kind, i.e., new modifications of an element, by scientifically planned, kinetically controlled syntheses. This was indeed possible. These new preparative methods enabled us to synthesize the thermodynamically unstable ring molecules $S_5(?)$, S_6, S_7, S_9, S_{10}, S_{11}, S_{12}, S_{18} (two allotropic forms), S_{20}, and $S_{24}(??)$. This chapter deals qualitatively with stereochemical aspects of some even-numbered (proper) rings in a general manner. Two review articles describe the synthetic routes as well as some properties of the new sulfur allotropes (2, 3).

Cyclodecasulfur, S_{10}, forms light yellow needles. It is rather unstable towards light as well as towards traces of bases and can only be stored for some time at temperatures below $0°C$. X-rays destroy it at room temperature within a few minutes. At $\sim -30°C$ the compound obviously is stable towards x-rays. A structure determination of such very sensitive crystals is now under way.

A rather big surprise was the first synthesis of cyclododecasulfur. In 1949, Pauling published a very important paper for sulfur chemistry (4). In this paper a mathematical correlation between bond distance, bond angle, dihedral angle, and the stability of staggered sulfur rings was developed. On the basis of his considerations and calculations, Pauling predicted for the still-hypothetical 12-membered sulfur ring an even higher ring strain and thus greater instability than in the extremely sensitive and labile cyclohexasulfur. This prediction proved to be wrong. Cyclododecasulfur, S_{12}, crystallizes in pale yellow needles. It does form regular solutions in CS_2, but its solubility is surprisingly low—less than 1% of that of S_8. The biggest surprise, however, was the thermal stability as well as the stability towards light. This allotrope of elemental sulfur may be stored in light at room temperature for many years without decomposition. Furthermore, whereas the most stable form of sulfur, S_8, partly decomposes with ring opening and thus seemingly melts at $\sim 119°C$, the new sulfur modification may be heated up to $148°C$. At this temperature it melts and decomposes, forming the usual and very complex mixture of molten sulfur. This unexpected stability and solubility made it possible to separate the new sulfur form from a large excess of S_8. Today we know that S_{12} is formed in the course of different reactions besides our original synthetic redox reaction between sulfanes and chlorosulfanes, for instance during the thermal decomposition of Steudel's new cyclooctasulfur-monoxide, S_8O, (5) or also by the photo-induced decomposition of S_6. In fact, we could isolate S_{12} from ordinary cooled sulfur melts after preheating S_8 to $\sim 120°-130°C$. This latter finding throws new light on the composition of sulfur melts at rather low temperatures, and the present interpretations urgently require reinvestigation.

Structure determination revealed reasons for the astonishing discrepancy between Pauling's prediction and our experimental findings. Pauling's considerations and mathematical correlations between ring strain and structural parameters in staggered sulfur rings was in principle correct and is still valid and very valuable. The x-ray determination revealed a highly symmetrical 12-membered ring (Figure 8). In this ring, however, the atoms are not arranged in two planes as is the case in Pauling's hypothesis, but in three planes; six atoms form a regular hexagon as the middle plane, three atoms are in a second plane above, and the other three in the third plane below the middle plane. With this conformation bond parameters are optimized and are similar to those found in the stable S_8.

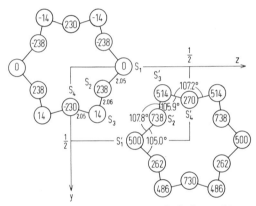

Figure 8.   Structure of cyclododecasulfur

A short time after the synthesis and structure determination of the 12-membered ring and without knowing of this synthesis, Tuinstra, as a consequence of very interesting and rather sophisticated considerations and calculations, predicted a sufficient stability at room temperature and the absolutely correct and precise conformation for the still-hypothetical S_{12} ring (6).

Tuinstra published his "Structural Aspects of the Allotropy of Sulfur and the Other Divalent Elements" in 1967. They are not discussed here in detail but are sketched only very roughly. In these considerations all excited or ionized states of the molecules are excluded, and only molecules in which all bonds are equal are taken into consideration—that is molecules with identical bond length, angles, and dihedral angles. Still the number of conformations for a distinct number of atoms remains large.

With respect to its three predecessors, any atom in a chain can make a choice from two different positions (as already mentioned earlier), namely the fourth atom initiates with its three predecessors (1) a right-handed or (2) a left-handed screw. If n is the number of atoms forming a molecule, the number of possible different molecular conformations is now $2^{(n-3)}$ ($n > 2$). With respect to only crystalline modifications, we can exclude all configurations in which the successive choices (1) and (2) are distributed at random; that means that only those configurations will be considered in which a distinct sequence of choices (1) and (2) is indefinitely repeated (that is, of course, also all ring molecules of that type).

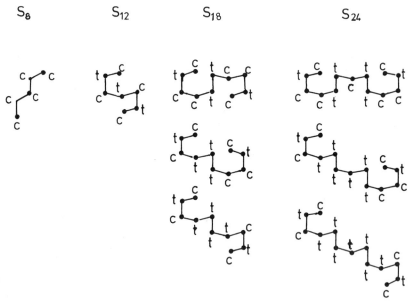

Figure 9. Motifs in sulfur rings. Only projection from the side is shown. Because of the plane of symmetry, corresponding atoms show up only once.

Because non-bonded atoms do not penetrate, a geometric model has much more effective space than the real molecules have. This important restricted volume limitation makes an elegant mathematical formulation impossible and was taken into account by Tuinstra in his final calculations.

The repeating unit is a distinct sequence of choices from (1) and (2); in the molecule this sequence corresponds to a row of successive atoms, the relative positions of which are fixed if bond and dihedral angles are kept constant. Such a repeating sequence of atoms is called a motif (or module). In general, a motif will coincide with its successor in the molecular string after a translation combined with a rotation in space. This implies that in general these molecular conformations are

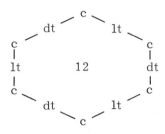

Figure 10.   Motif in S_{12}

helical. In this argument, as mentioned earlier, ring molecules may be considered as special helical molecules with pitch zero. Tuinstra differentiates between motifs leading to proper rings and other motifs leading to helical molecules including what he calls improper rings.

Going from geometric models to possible real molecular models requires rather complicated calculations. In doing this, Tuinstra took into account the bond parameters from experimental data, repulsion of next nearest neighbors, volume effect, etc. Because of the great calculation time, only the conformations of molecules with between four and 14 atoms have been postulated, including the conformation of S_{12}.

In the long-known S_8 (Figure 9) the very simple motif, cis–cis, is repeated four times. In the six-membered S_6, the same motif is repeated, but only three times. In the case of the surprisingly stable S_{12}, the more complicated motif, trans–cis–trans–cis, is repeated three times. More exactly within this sequence we have a regular change between dt and lt (Figure 10).

The next surprisingly stable sulfur ring synthesized was cyclooctadecasulfur, S_{18}. The rather orthorhombic crystals are sparingly soluble in CS_2 and are thermally stable up to the melting point at 128°. Again the stability against heat and also against light, is easy to understand since the bond parameters are similar to S_8. Figure 11 shows the structure of the molecule.

If we take three successive atoms out of a S_{12}-ring, then the remaining fragment of nine atoms combines easily with a second one to the new sulfur allotrope. This is the first example in our consideration, where the word ring is not really correct in the narrow sense of a circular molecule. We can look at molecules like the S_{18} as parallel helices linked

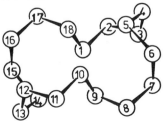

Figure 11.   Structure of cyclooctadecasulfur. Sulfur–sulfur distance ≈ 2.06 *Å; bond angle* $\beta \approx 106°$; *dihedral angle* $\gamma \approx 85°$.

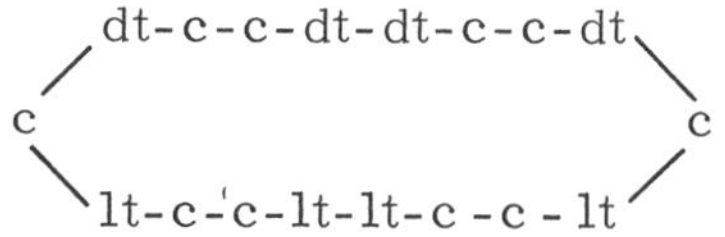

$$\begin{array}{c}
\text{dt-c-c-dt-dt-c-c-dt} \\
/ \qquad\qquad\qquad\qquad\qquad \backslash \\
\text{c} \qquad\qquad\qquad\qquad\qquad \text{c} \\
\backslash \qquad\qquad\qquad\qquad\qquad / \\
\text{1t-c-c-1t-1t-c -c - 1t}
\end{array}$$

Figure 12. Motif in S_{18}

on the end. This results from the more complicated motif, trans–cis–cis–trans–trans–cis–cis–trans–cis, that is repeated twice. In one side of the "ring" we have a left-handed helix, on the other side a right-handed one; the motif in detail is: *d*-trans–cis–cis–*d*-trans–*d*-trans–cis–cis–*d*-trans–cis–*l*-trans–cis–cis–*l*-trans–*l*-trans–cis–cis–*l*-trans–cis (Figure 12).

Recently a second modification of S_{18} has been found with the twice-repeated motif, cis–cis–trans–cis–trans–cis–trans–cis, again once *l*-trans and once *d*-trans (Figure 13) (7). The atoms lie on two planes nearly parallel to each other. The extreme two atoms of the elongated ring are in a trans position. Again, the ring is built up with the help of helices. All the ring parameters are similar to the hitherto discussed even-numbered sulfur rings, which agrees with the experimentally observed stability against light and heat.

The same also holds true for still another new sulfur allotrope, cyclo-icosasulfur, S_{20}. The orthorhombic crystals are stable up to $> 121°C$, but significantly less so than the two forms of S_{18} and S_{12}. The structure is shown in Figure 14. Again the parameters are similar to those in the aforementioned rings, but there are significant differences between the single values of γ. This, in connection with an abnormally long bond between the atoms 10 and 10' (2.104 Å) may be attributed to a closing position of the ring. The motif (Figure 15), cis–*l*-trans–*l*-trans–*l*-trans–cis, is repeated four times.

Future Work

Figure 9 shows some possible motifs and atomic arrangements of a ring size we consider very stable again, that is S_{24}. At the moment, however, this is rather speculative. We could isolate light yellow needles, stable up to $\sim 150°C$ which we think to be a 24-membered ring, but the structure determination has yet to be worked out.

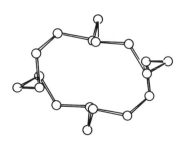

Figure 13. Structure of the second form of S_{18}

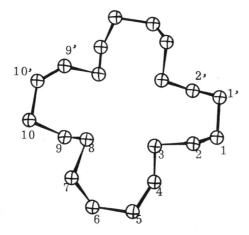

Figure 14. Structure of cycloicosasulfur.
Sulfur–sulfur distance ≈ 2.04 Å; bond
angle β ≈ 106°; dihedral angle γ ≈ 85°.

S_{10} and all the hitherto-known odd-numbered rings (S_7, S_9, S_{11}) can not be described with the help of regular motifs. The instability of those allotropes reflects the high ring strain of the molecules.

A very speculative consideration may be mentioned at this point: one can think of a large number of rather stable neutral sulfur rings composed of parallel helices linked end on. With normal bond distances, bond angles, and torsion angles they have the composition S_{6+n6} (S_{12}, S_{18}, S_{24}, S_{30}, S_{36} . . .). These theoretical molecules might well be major constituents of sulfur melts and of insoluble sulfur.

In this connection, a very recent paper by Allinger (8) should be mentioned. A force field has been developed to permit molecular mechanics calculations on various molecular structures of elemental sulfur. The conformational characteristics of sulfur rings containing five to 12, 14, 16, 18, or 20 sulfur atoms have been examined. Comparison with experimental data is made in all cases where such data exist, and predictions are made for other cases. Ab initio molecular orbital calculations using an STO-3G basis set were carried out for cyclohexasulfur and are consistent with the molecular mechanics calculations in indicating that the chair and the twist forms are two stable conformations, with the chair about 15 kcal/mol more stable while the boat (C_{2v}) is a twist rotational transition state. Calculation of possible conformations of the pro-

c-1t-1t-1t-c-c-1t-1t-1t-c

| | |
| | |

c-1t-1t-1t-c-c-1t-1t-1t-c   *Figure 15. Motif in S_{20}*

posed ring series S_{6+n6} by Allinger's method will offer a challenge to experimental chemists.

The very complicated sulfur melt can no longer be explained by the equilibria:

$$nS_8 \rightleftharpoons n \cdot S_8 \cdot \rightleftharpoons S_{8n}$$

There are convincing facts that in the melt at any given temperature there are also molecules other than S_8 and multiples of S_8. Some time ago we could isolate S_{12} from molten sulfur. Only a few months ago Steudel (9) reported very interesting and promising new studies on quenched melts by Raman and IR spectroscopy. He found S_6 and S_7 in significant amounts. Other not-yet-identified lines indicate the probable presence of S_5, S_9, and/or S_{11}.

Meyer (3) showed some time ago that the color of hot sulfur melts is caused mainly by the presence of S_3 and S_4. In this connection we should also mention recent sophisticated studies by Block and co-workers (10). Sulfur molecules S_x with 2–22 sulfur atoms have been desorbed from a condensed sulfur layer on a tungsten field emitter of a field ionization time-of-flight mass spectrometer. The condensed sulfur layer is in a highly mobile liquid-like steady state. The observation of these large sulfur molecules is important to the current models of liquid sulfur.

At the present time, the scientific basis of practical applications of elemental sulfur unfortunately is not very strong in many respects. The situation has been improved, however, in the past 10 years, but much more work is needed so that we can influence the properties of elemental sulfur to meet the requirements of different practical applications.

Literature Cited

1. Steudel, R., *Angew. Chem.* (1975) **87**, 683.
2. Schmidt, M., *Angew. Chem.* (1973) **85**, 474.
3. Meyer, B., *Chem. Rev.* (1976) **76**, 367.
4. Pauling, L., *Proc. Natl. Acad. Sci. U.S.* (1949) **35**, 495.
5. Steudel, R., *Angew. Chem.* (1972) **84**, 344.
6. Tuinstra, F., "Structural Aspects of the Allotropy of Sulfur and the Other Divalent Elements," Uitgeverij Waltmann, Delft, 1967.
7. Debaerdemaeker, T., Kutoglu, A., *Cryst. Struct. Commun.* (1974) **3**, 611.
8. Koa, J., Allinger, N. L., *Inorg. Chem.* (1977) **16**, 35.
9. Steudel, R., *Angew. Chem.* (1977) **16** (2), 112.
10. Cocke, D. L., Abend, G., Block, J. H., *J. Phys. Chem.* (1976) **80**, 524.

RECEIVED April 22, 1977.

2

Preparation and Properties of Modified Sulfur Systems

L. BLIGHT, B. R. CURRELL, B. J. NASH,
R. A. M. SCOTT, and C. STILLO

Thames Polytechnic, Wellington Street, Woolwich, London SE18 6PF, England

Materials prepared by modifying sulfur with dicyclopenta-diene or styrene consist of unreacted sulfur and polysulfides. Dicyclopentadiene reacts with sulfur to give, initially, low-molecular-weight polysulfides (e.g., a pentasulfide) which convert to high-molecular-weight polymers at 140°C. Styrene initially gives high-molecular-weight polymers which depolymerize. With dicyclopentadiene N,N,N',N'-tetra-methylethylenediamine catalyzes the formation of low- and inhibits the formation of high-molecular-weight material and therefore offers promise as a viscosity control agent. The mechanical properties of modified sulfur vary with time after preparation. Modifiers can stop or reduce the embrittlement of elemental sulfur. The tensile strength of polypropylene and glass–fiber fabrics may be increased by up to 87 and 104%, respectively, on impregnation with sulfur materials.

Elemental sulfur has been proposed (*1, 2*) for a wide range of applications in the civil engineering field. In virtually all of these applications it has been necessary to modify the sulfur with additives designed to stop the embrittlement which occurs with pure elemental sulfur. Thus if pure elemental sulfur is heated to 140°C and then cooled to ambient temperature, monoclinic sulfur (S_β) is instantaneously formed (*3*), followed by a reversion to orthorhombic sulfur (S_α) which is almost complete in about 20 hr (Figure 1). Many additives have been proposed to modify elemental sulfur, nearly all of which fall under the heading of polymeric polysulfides or, alternatively, substances which may react with elemental sulfur to give in situ formation of polymeric polysulfides. In a previous paper (*3*), we reviewed the various substances used as addi-

0-8412-0391-1/78/33-165-013$05.00/0

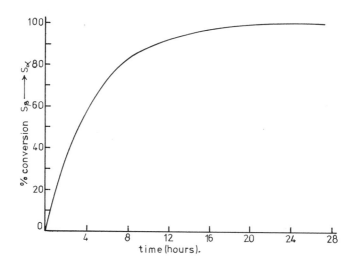

Figure 1. Reversion rate of monoclinic sulfur (S_β) to ortho-rhombic sulfur (S_α) at ambient temperature

tives and compared quantitatively their effectiveness in retarding the crystallization of elemental sulfur. The additives studied were a range of olefins and certain Thiokols (polymeric polysulfides). The results are summarized in Table I; each of these additives (25% w/w sulfur) was mixed with elemental sulfur and heated at 140°C for 3 hr. In each case the resulting material was a mixture of polysulfides and unreacted elemental sulfur. These additives have a significant effect in reducing the formation of orthorhombic sulfur, the olefins (except styrene) being

Table I. Composition of Modified Sulfur Materials[a]

	Poly-sulfides (%)	Unreacted Free Sulfur (%)			
		S_α	S_β	S_8 (liq.)	Total S
D (+) Limonene	40.0	0.0	36.0	24.0	60.0
Dicyclopentadiene	45.6	0.0	30.5	23.9	54.4
Alloocimene	39.6	0.0	45.8	14.6	60.4
Myrcene	53.6	0.0	31.2	15.2	46.4
Cycloocta-1,3-diene	47.9	0.0	40.0	12.1	52.1
Cyclododeca-1,5,9-triene	53.7	0.0	38.1	8.2	46.3
Styrene	26.0	60.0	9.0	5.0	74.0
Thiokol LP-31	50.8	17.4	31.8	0.0	49.2
Thiokol LP-32	47.5	24.2	28.3	0.0	52.5
Thiokol LP-33	39.8	36.0	24.2	0.0	60.2

[a] After storage for 18 mo at ambient temperature. Percentage figures given each refer to the percentage of the total composition.

particularly effective with no orthorhombic sulfur formed, even after 18 mo.

This chapter reports the study of the chemical composition of the polysulfide fraction in dicyclopentadiene and styrene modified materials, the mechanical properties of modified materials, and their use in the preparation of composite materials by the impregnation of polypropylene and glass–fiber fabrics.

Chemistry of Sulfur Modification

Sulfur Modified with Dicyclopentadiene. The beneficial use of dicyclopentadiene to modify sulfur has been reported by a number of workers including Currell et al. (3), Sullivan et al. (4), and also Diehl (5). Currell et al. showed that the interaction of dicyclopentadiene and elemental sulfur at 140°C gives a mixture of polysulfides and free elemental sulfur which, even after standing for 18 mo, is held as a mixture of presumably monoclinic and noncrystalline sulfur. Sullivan et al. reported that the minimum concentration of dicyclopentadiene required to stop permanently the embrittlement of elemental sulfur is 13% if the reaction temperature is less than 140°C and only 6% if the reaction temperature is greater than 140°C. Presumably the polysulfide reaction products form a solid solution with the unreacted sulfur from which orthorhombic sulfur cannot crystallize.

Table II summarizes the initial results of a further study on the modification of sulfur by dicyclopentadiene at 140°C. The efficiency of dicyclopentadiene in inhibiting the formation of orthorhombic sulfur is

Table II. Composition of Sulfur Materials Modified by Dicyclopentadiene[a]

Modifier Loading (%)	Heating Time (hr)	Poly- sulfides (%)	Unreacted Free Sulfur (%)				Fraction Insol. in CS_2 (%)
			S_α	S_β	S_8 (liq)	Total S	
5	3	13.4	54	5	27.6	86.6	11.4
10	3	25.5	0	38	36.5	74.5	13.4
25	1	[c]	[c]	[c]	[c]	[c]	[c]
25[b]	1	31.5	0	21	48.4	69.4	0
25	3	54.7	0	17	28.3	45.3	15.1
25[b]	3	48.7	0	30	21.3	51.3	0
25	20	68.7	0	0	31.3	31.3	56.6
40	10	64.3	0	0	35.7	35.7	0

[a] After storage for 18 mo at ambient temperature. Percentage figures given each refer to percentage of total composition.
[b] N,N,N',N'-tetramethylethylenediamine (0.1%) added.
[c] Incomplete reaction, olefin still present.

again demonstrated. The fact that orthorhombic sulfur is formed with a loading of 5%, and not with a 10% loading, is in broad agreement with the minimum necessary levels reported by Sullivan et al.

The amount of polysulfides formed increases with the amount of dicyclopentadiene added at 140°C. Thus with a 25% loading after 1 hr, the reaction is incomplete, with unreacted olefin still present; after 3 hr, 54.7% of the final product is polysulfides; and after 20 hr, 68.7% is polysulfides. Also, as reaction time proceeds, the molecular weight of the polysulfide increases.

The reaction products were examined initially by extraction with carbon disulfide to give, in many cases, a fraction soluble in carbon disulfide. In every case all the unreacted free sulfur reported in Table II was soluble in carbon disulfide, indicating that it is non-polymeric and presumably ring material. The amount of insoluble material formed increases with reaction time. This insoluble fraction is a high-molecular-weight, cross-linked material, which swells in carbon disulfide and has an elemental analysis which corresponds to $C_{10}H_{12}S_{11}$. Further examination has not been possible because of its intractable nature.

Those reactions in which N,N,N',N'-tetramethylethylenediamine was added gave no formation of insoluble material. In addition, the rate of formation of the soluble polysulfides was increased as shown not only by the amount of these polysulfides formed but also by their molecular weight distributions (see below). With a 25% loading after 1 hr, reaction is incomplete, with unreacted olefin still present. But, if the diamine is used, all the olefin reacts after 1 hr, and the resultant material contains 31.5% soluble polysulfides.

We can confirm the reports of Sullivan et al. that the reaction between dicyclopentadiene and sulfur is exothermic. If the temperature rises above 150°C, the extreme viscosity increase causes the mixture to become almost solid and the reaction difficult to control. Diehl (5) and Bordoloi and Pearce (6, 7) have reported quantitative studies of these viscosity changes. They show that there are large viscosity increases as the amount of dicyclopentadiene added, reaction temperature, and reaction time are increased. Our results show that these increases in viscosity are caused by the formation of high-molecular-weight polysulfides.

Diehl has also investigated the compressive strength of concrete, prepared by using sulfur and modified with dicyclopentadiene, as a binder for a standard aggregate (standard sand to DIN 1164). His results show that the compressive strength of the concrete is very dependent on the loading of dicyclopentadiene and also on the reaction time and temperature. This dependence of the chemical constitution and properties on dicyclopentadiene loading, reaction time, and temperature is obviously a severe disadvantage in any commercial application. In such applications melts may need to be kept at elevated temperatures for different periods of time, and also, in the field accurate temperature control may not be as easy as in the laboratory. In addition the possibility of the accidental formation of very high-viscosity material may not only make

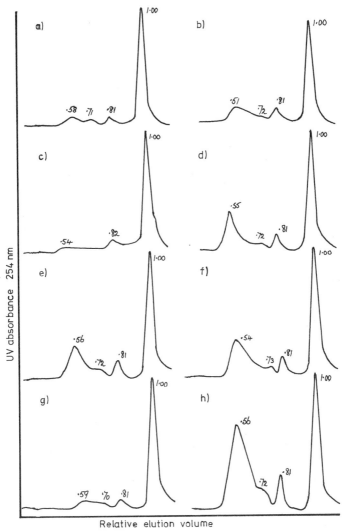

Figure 2. Gel permeation chromatograms of the carbon disulfide-soluble fraction from dicyclopentadiene-modified sulfur materials. Samples are identified by percentage of dicyclopentadiene used (w/w sulfur) and heating at 140°C. (a) 5%, 3 hr; (b) 10%, 3 hr; (c) 25%, 1 hr; (d) 25%, 1 hr; (e) 25%, 3 hr; (f) 25%, 3 hr; (g) 25%, 20 hr; (h) 40%, 10 hr. In the case of (d) and (f) N,N,N'N'-tetramethylethylenediamine was added to the reaction mixture.

equipment difficult to clean but impossible to re-use. Our observations on the effects of N,N,N',N'-tetramethylethylenediamine in stopping the formation of high-viscosity material and catalyzing the formation of lower-molecular-weight polysulfides may lead to the development of material whose constitution and properties are much more tolerant of changes in reaction conditions.

The nature of the soluble polysulfides has been investigated in detail. Gel permeation chromatography using chloroform as a solvent indicates fractions at V_{RE} 1.00, 0.81, 0.72, and 0.56. 1.00 is elemental sulfur, and the lower the V_{RE} figure, the higher the molecular weight. These curves for different reaction products are given in Figure 2. The relative proportions of the higher-molecular-weight fractions increase with respect to both reaction time and the amount of dicyclopentadiene added (cf., increases in viscosity). The catalytic effect of N,N,N',N'-tetramethylethylenediamine can be seen by comparing c with d and e with f.

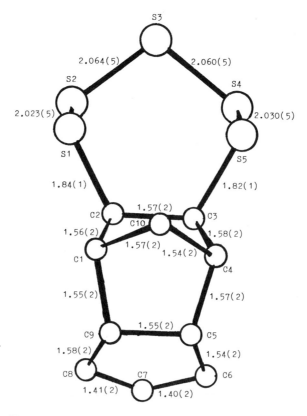

Figure 3. Structure obtained by x-ray diffraction of cyclic pentasulfide formed by interaction of sulfur and dicyclopentadiene

The various fractions have been separated and identified. The material corresponding to fraction V_{RE} 0.81 was obtained by extracting the 25% (3 hr) reaction product with hot chloroform followed by preparative gel permeation chromatography to give, after removal of solvent under reduced pressure, a yellow glassy solid (Found: C, 45.7, H, 4.7; S, 52.7; $\overline{M_n}$ 299. Calculated for $C_{10}H_{12}S_5$: C, 41.1; H, 4.1; S, 54.8; $\overline{M_n}$ 292). This product is a mixture of cyclic tri- and pentasulfide, each formed by adding sulfur across the bicycloheptenyl double bond. The evidence for this structure rests on mass, infra-red, and nuclear magnetic resonance spectroscopy and on the preparation of model compounds using active sulfuration.

The structure of the pentasulfide was finally confirmed by x-ray crystallography (Figure 3). We have not been able to resolve the position of the double bond as either C_6–C_7 or C_7–C_8. This may be caused by delocalization effects, pseudo-symmetry, or the fact that the product is actually a mixture of the two isomers.

The isolation of fraction V_{RE} 0.58 was accomplished by adsorption chromatography of the 40% (10 hr) reaction product. Analysis (Found: C, 33.1; H, 3.5; S, 63.0; $\overline{M_n}$ 3390) is consistent with the average structure given below.

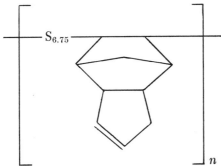

It is possible that this polymer is formed by ring-opening polymerization of the cyclic pentasulfide or other cyclic polysulfides. In a separate experiment, heating the trisulfide gave a similar chain polymer. We have not been able to separate the fraction $V_{RE} = 0.72$ which was formed in much smaller quantities than other fractions.

There is a strong possibility that in the examination of both these dicyclopentadiene products and those of styrene to be described, the process of extraction and examination will cause some transformation of the products because of the lability of polysulfide linkages. Thus these results should be treated with caution, but the authors believe the results give a good indication of the type of products formed.

Table III. Composition of Sulfur Materials Modified by Styrene[a]

Modifier Loading (%)	Heating Time (hr)	Poly-sulfides (%)	Unreacted Free Sulfur (%)			
			S_α	S_β	S_8 (liq)	Total S
5	3	13.2	64	0	22.8	86.8
10	3	21.8	0	53	25.2	78.2
25	0.75	43.4	0	44	12.6	56.6
25	2.50	46.5	0	35	18.5	53.5
25	13.5	42.8	0	37	20.2	57.2
25	200	27.7	72.3	0	0	72.3
66	3	83.4	0	0	16.6	16.6
66	6	86.9	0	0	13.1	13.1

[a] After storage for 18 mo at ambient temperature. Percentage figures given each refer to percentage of total composition.

Sulfur Modified with Styrene. Table III summarizes the results of the examination of the modification of sulfur with styrene (5, 10, 25, and 66% w/w sulfur). As with the dicyclopentadiene reaction, the reaction product is a mixture of unreacted free sulfur and polysulfides. In this series of experiments orthorhombic sulfur was formed with 5% styrene only. Our earlier work indicated that orthorhombic sulfur is formed with the use of 25% styrene. We can offer no firm explanation for this discrepancy, but in this earlier work much higher-molecular-weight polysulfides were formed, and the differences may result from a critical dependence on reaction time and temperature. The reaction of sulfur with styrene is exothermic; a temperature rise being observed about 30 min after addition. This rise is associated with a viscosity increase. However, in contrast to the dicyclopentadiene reaction, the viscosity decreased within 2 hr, and the reaction mixture was stirred easily thereafter. Each reaction product is completely soluble in carbon disulfide and also, with one exception, in chloroform. Thus after 45 min at 140°C, the 25% styrene reaction product results in a product, 39% of which is insoluble in chloroform. This insoluble fraction comprises almost the whole of the polysulfide fraction. But all longer heating times produce a completely soluble product. Thus the reaction produces initial formation of high-molecular-weight material (*see also* Ref. 3), which then at 140°C progressively depolymerizes. This effect is shown also in the viscosity changes and in the molecular weight distribution of the chloroform-

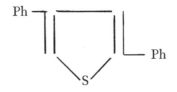

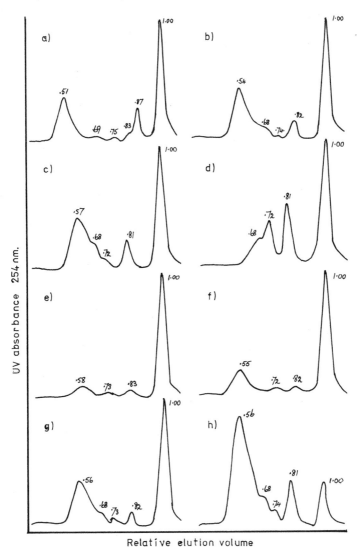

Relative elution volume

Figure 4. Gel permeation chromatograms of the styrene-modified sulfur materials. Samples are identified by percentage of styrene used (w/w sulfur) and heating time at 140°C. (a) 25%, 0.75 hr; (b) 25%, 2.5 hr; (c) 25%, 13.5 hr; (d) 25%, 200 hr; (e) 5%, 3 hr; (f) 10%, 3 hr; (g) 25%, 3 hr; (h) 66%, 3 hr. In the case of (g), N,N,N'N'-tetramethylethylenediamine was added to the reaction mixture.

soluble components. Gel permeation chromatograms of the chloroform-soluble components are given in Figure 4. The two principal components have peaks at V_{RE} 0.82 and at 0.56. As reaction time is increased with a 25% loading, the proportion of the lower-molecular-weight component

(0.82) increases. These components have been separated and the material corresponding to V_{RE} 0.82 identified as 2,4-diphenylthiophene. The identification is based on elemental analysis, melting point, and infrared, mass, and nuclear magnetic resonance spectroscopy.

All the other reaction products appear to be chain styrene polysulfides of different molecular weights. The two oligomers I and II have been obtained from the fraction V_{RE} approx. 0.73, and fully characterized by elemental analysis, molecular weight measurement, and various spectroscopic techniques.

On standing for 1 mo under ambient conditions, II partially decomposed to 2,4-diphenylthiopene. A fraction corresponding to V_{RE} 0.56 was isolated by fractional precipitation of a chloroform solution of the 66% (6 hr) product with petroleum ether (60°–80°C). The orange polysulfides amounting to 63% of the total product analyzed according to the molecular formula $(C_8H_8S_{4.5})_n$. Determination of molecular weight ($\overline{M}_n$ 1138) gives a degree of polymerization of approximately 5, and deter-

I

II

III

mination of the sulfur rank gives an average sulfur chain length of 4.4. $LiAlH_4$ hydrogenation gave phenylethane-1,2-dithiol and 1-phenylethane-thiol suggesting an average structure (III).

Reactions were also attempted in the presence of N,N,N',N'-tetra-methylethylenediamine. Unlike the dicyclopentadiene examples these reactions gave copious formation of hydrogen sulfide.

Mechanical Properties of Modified Sulfur Materials

As we reported previously (3), the crystalline sulfur content of modified sulfur materials increases after preparation; at least a month is needed to reach a stable level. We have now begun to study the mechanical properties of these materials and their variation with time. This study is still in progress, and Figures 5, 6, 7, and 8 contain the preliminary results. Each of these materials was made by heating the sulfur for 3 hr at 140°C with the stated additive except for myrcene sulfur, which was heated for 48 min at 140°C since the viscosity rose to prevent adequate stirring. The points marked X in the hardness curves (Figure 5) indicate the point after which the specimens failed on test. After this point the hardness value is merely a measure of the brittle fracture strength of the material. The results agree with the previous results of the amounts of additives needed to prevent embrittlement of sulfur. Thus pure sulfur is brittle in 1 hr, sulfur modified by 5 and 10% dicyclopenta-diene in a few days, 5% styrene in one day; but no embrittlement has been seen yet in 25% dicyclopentadiene, 25% myrcene, 25% styrene, and 10% styrene.

Compressive strength (Figure 6) of sulfur has a very high initial figure falling to a level at about 7000 kNm^{-2}. The modified materials have much lower values, which with the lower levels of additive (5 and 10% styrene), increase rapidly after approximately two weeks, presum-

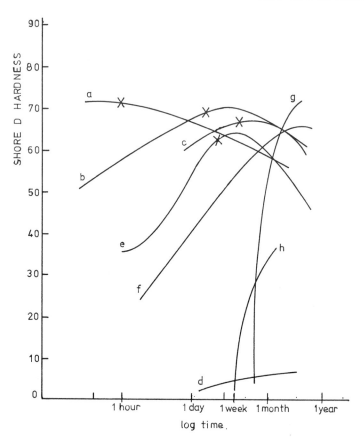

Figure 5. Shore D hardness vs. log time plots of modified sulfur materials. Samples are identified by percentage of modifier used (w/w sulfur) and heating time at 140°C. The onset of sample shattering is indicated by X. (a) Sulfur; (b) 5% dicyclopentadiene, 3 hr; (c) 10% dicyclopentadiene, 3 hr; (d) 25% dicyclopentadiene, 3 hr; (e) 5% styrene, 3 hr; (f) 10% styrene, 3 hr; (g) 25% styrene, 3 hr; (h) 25% myrcene, 0.8 hr.

ably from formation of crystalline sulfur. The values for 10 and 25% dicyclopentadiene were too small to be measured until 1 mo.

Changes in ultimate tensile strength (UTS) (Figure 7) are less easy to rationalize. Presumably one controlling factor is the crystallization of sulfur, but other factors appear to be more important. The materials containing 25% styrene and 25% myrcene show no strength until 1 week; this is then followed by a rapid increase. The elongation figures (Figure 8) obtained in these tensile tests show similar anomalies. Sulfur shows little elongation, but the modified materials have initial elongation values which increase with the level of added modifier.

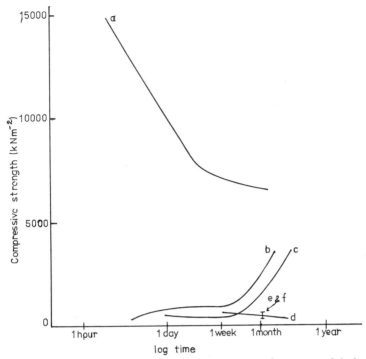

Figure 6. Compressive strength vs. log time plots of modified sulfur materials. Samples are identified by percentage of modifier used (w/w sulfur) and heating time. (a) Sulfur; (b) 10% styrene, 3 hr; (c) 5% styrene, 3 hr; (d) 25% myrcene, 0.8 hr; (e) 10% dicyclopentadiene, 3 hr; (f) 25% dicyclopentadiene, 3 hr.

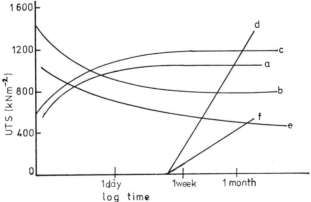

Figure 7. Ultimate tensile strength vs. log time plots of modified sulfur materials. Samples are identified by percentage of modifier used (w/w sulfur) and heating time. (a) Sulfur, 3 hr; (b) 5% styrene, 3 hr; (c) 10% styrene, 3 hr; (d) 25% styrene, 3 hr; (e) 5% dicyclopentadiene, 3 hr; (f) 25% myrcene, 0.8 hr.

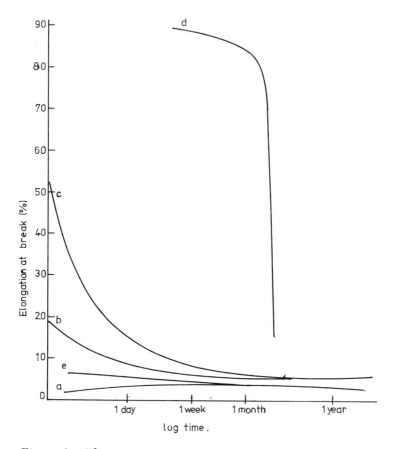

Figure 8. Elongation at break vs. log time of modified sulfur materials. Samples are identified by percentage of modifier used (w/w sulfur) and heating time. (a) Sulfur; (b) 5% styrene, 3 hr; (c) 10% styrene, 3 hr; (d) 25% styrene, 3 hr; (e) 5% dicyclopenta-diene, 3 hr.

Aging of these materials in addition to changes in sulfur crystallinity may involve changes in the nature of the polysulfide fractions. For example, one styrene polysulfide has an average sulfur rank of 6.75. This is extremely high, and it can be expected to be very unstable. Tetrasulfides and above are unstable and decompose, forming polysulfides with lower sulfur rank and elemental sulfur. Future work will explore this possibility.

Sulfur–Fiber Composites

In association with the Lambeg Industrial Research Association we have studied the preparation and properties of sulfur-based fiber composites. The ultimate objective is to prepare modified sulfur/polypropylene fiber composites which are flexible and which possess suitable mechanical properties for certain civil engineering applications.

Sulfur–fiber composites have been prepared and studied by a number of workers. Rods have been prepared (8) by passing parallel orientated glass strands through molten sulfur under vacuum to form reinforced bars 150 mm long and 6 mm in diameter. Glass–fiber mats have been impregnated with sulfur also, the mat being covered with powdered elemental sulfur and then placed in an oven at 185°C for 1.5 hr so that polymeric sulfur is formed.

Dale and Ludwig (9) also investigated the use of glass–fiber mats and had difficulty in wetting the fibers with molten sulfur, but including a modifier (e.g., a terpene) improved the wetting ability and also increased the strength of the composite.

Sulfur, modified in its structure so as to become plastic, presents other problems as it can be a very ductile material. Reinforcement in the sense of load sharing may be effected by incorporating relatively ductile thermoplastic fibers. There is a commercial interest in using polypropylene fibers, and Bennett (10, 11) has patented a process whereby woven polypropylene fibers are impregnated with asphalt or bitumen and then are coated on at least one side with a sulfur composition. The plasticized sulfur coating sticks to the base and makes the laminate flexible, weather resistant, and light reflecting.

Mechanical Properties of Sulfur–Fiber Composites. Impregnation was obtained by simply dipping the fabric into the molten mixture at 130°C; the impregnated strip was then removed and surplus liquid allowed to drip off. Better controlled impregnation was carried out by hot-pressing the fabric and granules of the modified sulfur at 130°C. The pressure applied, approximately 1.0 Nmm^{-2}, is not significant as the sheets were pressed into a mold to a fixed thickness of 1.5 mm. Table IV summarizes the tensile test results of impregnated polypropylene fabric, and Table V relates to similar results on impregnated glass fabric. The limits of error given are the standard deviations obtained from at least five tests on apparently identical impregnated strips. The loadings of modifier in the sulfur mixtures were 25% w/w unless otherwise stated. Results are given for each fabric impregnated with various modified sulfur materials, pure sulfur, and the unimpregnated fabric. Results are also given for an unimpregnated fabric which has been heat-treated at 130°C prior to testing.

Increases in the UTS of the polypropylene fabric by impregnation with sulfur materials ranged from 24 to 87%; the highest increases were obtained with dicyclopentadiene-modified sulfur materials. Merely heat-

Table IV. Mechanical Properties of

Reinforcement	Nature of Product	UTS (kNm⁻¹)	
		Warp	Weft
None	flexible	8.7 ± 1.1	9.9 ± 0.8
None[a]	flexible	10.1 ± 0.3	11.5 ± 0.2
Pure S	brittle	12.4 ± 1.5	14.6 ± 1.5
S/styrene	brittle	11.9 ± 0.1	12.3 ± 1.0
S/myrcene	flexible	12.9 ± 0.4	14.0 ± 0.8
S/alloocimene	flexible	13.0 ± 0.4	13.8 ± 0.9
S/cycloocta 1,3-diene	flexible	13.5	13.1
S/10% w/w DCPD	brittle	14.2	17.5
S/15% w/w DCPD	brittle	16.3	16.8
S/DCPD	brittle	13.5 ± 1.3	15.4 ± 3.0

[a] Heat-treated material.

Table V. Mechanical Properties

Reinforcement	Nature of Product	UTS (kNm⁻¹)	
		Warp	Weft
None	flexible	33.1 ± 2.5	30.5 ± 3.2
Pure S	brittle	41.4 ± 3.9	33.1 ± 2.0
S/styrene	brittle	57.8 ± 0.5	43.5 ± 2.8
S/myrcene	flexible	67.5 ± 1.2	52.3 ± 2.2
S/alloocimene	flexible	50.0 ± 5.7	55.0 ± 2.3
S/DCPD	brittle	62.7 ± 6.4	45.8 ± 3.2

ing polypropylene fabric to 130°C increases the UTS and elongation at break. This effect is thought to be caused by changes in the crystalline regions within the fabric.

A more significant comparison is therefore made between the impregnated fabrics and heat-treated fabric. Here the increase in UTS by impregnation ranges from 7 to 61%. Myrcene, alloocimene, and cyclo-octa-1,3-diene all produced flexible materials.

The strengthening effect of modified sulfur is greater with glass fabric than it is with polypropylene fabric probably because the modifiers used react with the chemical keying agents on the glass surface to provide a more strongly bonded composite pair. Improvements in the UTS by impregnation of glass fabric range from 43 to 104%. The most effective is myrcene which also gives a flexible product.

In conclusion, both woven polypropylene and glass–fiber fabric each give strong, flexible weatherproof sheet materials when impregnated by sulfur modified by myrcene, alloocimene, or cycloocta-1,3-diene. Typical applications for this type of material are in the fields of civil engineering

Impregnated Polypropylene Fabric

Extension at Break (%)		Young's Modulus (kNm^{-1})	
Warp	*Weft*	*Warp*	*Weft*
13.9 ± 1.2	17.4 ± 4.6	66.1 ± 8.0	67.4 ± 14
23.1 ± 0.1	29.0 ± 0.3	48.6 ± 1.6	43.3 ± 3.0
23.4 ± 3.6	25.8 ± 2.6	55.8 ± 12.2	58.1 ± 6.9
17.0 ± 0.2	24.5 ± 0.6	64.1 ± 2.8	48.7 ± 2.7
17.1 ± 2.6	23.7 ± 0.4	63.0 ± 1.4	52.4 ± 4.0
20.6 ± 0.1	31.5 ± 1.3	62.9 ± 3.3	46.3 ± 3.0
24.5	27.6	61.0	50.4
37.1	42.8	39.2	41.5
29.6	34.4	59.1	50.3
21.1 ± 1.9	26.7 ± 6.0	61.8 ± 9.6	50.2 ± 2.5

of Impregnated Glass Cloth

Extension at Break (%)		Young's Modulus (kNm^{-1})	
Warp	*Weft*	*Warp*	*Weft*
2.2 ± 0.4	2.4 ± 0.3	1.7 ± 0.2	1.3 ± 0.1
2.9 ± 0.4	2.5 ± 0.3	1.8 ± 0.1	1.5 ± 0.1
4.0 ± 0.5	3.2 ± 0.1	1.7 ± 0.1	1.6 ± 0.1
4.9 ± 0.5	5.1 ± 0.2	—	—
4.6 ± 0.4	5.7 ± 0.2	—	—
3.2 ± 0.1	3.9 ± 0.3	2.8 ± 0.4	1.5 ± 0.1

and building construction as roofing materials, damp-proof courses, and liners for ponds, canals, and irrigation ditches.

Experimental

The details of the purification of sulfur, the preparation of modified materials and of many of the analytical methods used are given in Ref. 3.

Analytical Gel Permeation Chromatography. A Waters system comprising constant flow rate pump (model M 6000) and differential uv (254 nm) and R.I. detectors was used. Waters μ-styragel columns (10^4, 10^3, 5×10^2, and 10^2 Å) were used in series for molecular size distributions, usually with chloroform as the eluting solvent. The elution values (V_{RE}) of the fractions are quoted relative to that of elemental sulfur.

Tensile Strength. A Hounsfield tensometer was used (extension rate 6.33 mm/min). Specimens were cast (140°C) into waisted tensile molds (length 92 mm $\times$ 0.5 mm $\times$ 0.6 mm in waisted section) and were tested using wedge-shaped grips with thin card packing between grips and specimen.

Compressive Strength. Specimens were cast (140°C) into polyethylene molds (8.2 mm diameter, 2.5 mm thickness), removed, and tested.

Composite Tensile Testing. Tests were according to principles of BS 2576 with reduced specimen dimensions (14.6 mm × 2.4 mm). Specimens were held in vice grips with corrugated pressure faces (gage length 80 mm).

Acknowledgment

The authors wish to thank the Sulphur Institute and also the Lambeg Industrial Research Association for financial support.

Literature Cited

1. "New Uses of Sulfur," Adv. Chem. Ser. (1974) **140**.
2. "New Uses for Sulfur and Pyrites," Madrid Symposium, The Sulphur Institute, 1976.
3. Currell, B. R., Williams, A. J., Mooney, A. J., Nash, B. J., Adv. Chem. Ser. (1974) **140**, 1.
4. Sullivan, T. A., McBee, W. C., Blue, D. D., Adv. Chem. Ser. (1974) **140**, 55.
5. Diehl, L., "New Uses for Sulfur and Pyrites," Madrid Symposium, The Sulphur Institute, 1976.
6. Bordoloi, B. K., Pearce, E. M., Sulphur Institute Internal Report.
7. Bordoloi, B. K., Pearce, E. M., Adv. Chem. Ser. (1978) **165**, 31.
8. Harris, R., U.S. Patent 3183143 (May 11th, 1965).
9. Dale, J. M., Ludwig, A. C., *Sulphur Inst. J.* (1969) **5**, 2.
10. Bennett, R., U.S. Patent 3619258 (December 22, 1971).
11. Bennett, R., U.S. Patent 3721578 (March 20, 1973).

Received April 22, 1977.

Plastic Sulfur Stabilization by Copolymerization of Sulfur with Dicyclopentadiene

BINOY K. BORDOLOI and ELI M. PEARCE

Department of Chemistry, Polytechnic Institute of New York,
Brooklyn, NY 11201

Liquid sulfur–dicyclopentadiene (DCP) solutions at 140°C undergo bulk copolymerization where the melt viscosity and surface tension of the solutions increase with time. A general melt viscosity equation: $\eta = \eta_o \, exp(aX^b t)$, at constant temperature, has been developed, where η is the viscosity at time t for an S_8–DCP feed composition of DCP mole fraction X; and η_o (in viscosity units), a (in time^{-1}), and b (a dimensionless number, $+$ ve for X $<$ 0.5 and $-$ve for X $>$ 0.5) are empirical constants. The structure of the sulfurated products has been analyzed by NMR. Sulfur noncrystallizable copolymeric compositions have been obtained as shown by thermal analysis (DSC). Dodecyl polysulfide is a viscosity suppressor and a plasticizer for the S_8-DCP system.

S ulfur is a readily available, high-purity and low-cost material (*1*). In 1975 in the western world 34.2 and 30.6 million tons of all forms of sulfur were produced and consumed, respectively (*2*). Fike (*1*) has pointed out some potential applicaitons of this material.

Sulfur is used directly in sulfur-additive solutions which are used as coatings (*3, 4, 5, 6*). Dicyclopentadiene is a potentially cheap material that can form copolymers with liquid sulfur. The patent literature reports applications of the sulfur–dicyclopentadiene system in sprayable coatings, rubber vulcanizates, etc. However, very little has been written on the chemistry, characterization, and properties of this system.

Chemistry of the Sulfur–Dicyclopentadiene Reaction

Bicyclo (2.2.1) heptene (norbornene), **I**, is a strained olefin. Dicyclopentadiene (DCP), **II**, is a substituted norbornene, and it has two stereo-

0-8412-0391-1/78/33-165-031$10.00/0

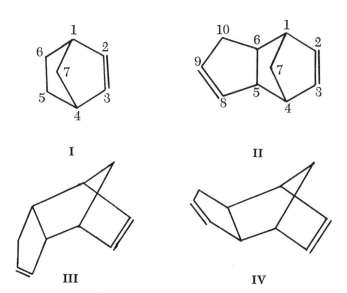

isomers, *endo*-DCP (**III**) and *exo*-DCP (**IV**). Commercially available DCP is almost exclusively the endo isomer formed from cyclopentadiene by dimerization according to the cis addition rule of Alder (7). However, the exo isomer is the thermodynamically more stable isomer (7). Co-polymerization experiments have shown that the norbornene unsaturation in *endo*-DCP is twice as reactive as the cyclopentene unsaturation, *exo*-DCP is twice as reactive as the *endo*-DCP, and *endo*-DCP is 7.5 times more reactive than propylene (7).

Sulfur–olefin reactions are categorized as low temperature reactions up to about 140°C and as high temperature reactions at over 140°C (8). Although a free-radical chain-growth mechanism was proposed initially for the low-temperature reactions, a cationic chain-growth mechanism appeared to have more experimental support at this time. However, these were not polymer-forming reaction conditions. High-temperature reactions are complex, and no unique mechanism has been proposed, although both free-radical and cationic mechanisms are expected to occur. We have studied the bulk copolymerization reaction of dicyclopentadiene at low-temperature reactions with liquid sulfur.

Dicyclopentadiene (boiling point for both the isomers is 170°C) is soluble in liquid sulfur at 140°C in all proportions, and the melt viscosity of the sulfur–DCP solution increases with time because of a copolymerization reaction. The DSC thermograms of sulfur–DCP mixtures are shown in Figure 1. The exothermic reaction clearly becomes evident with a feed of about 20% DCP with the exotherm starting at about 140°C.

The nature of certain sulfur–olefin reactions carried out in vacuum-sealed ampoules and taking equimolar compositions of purified materials

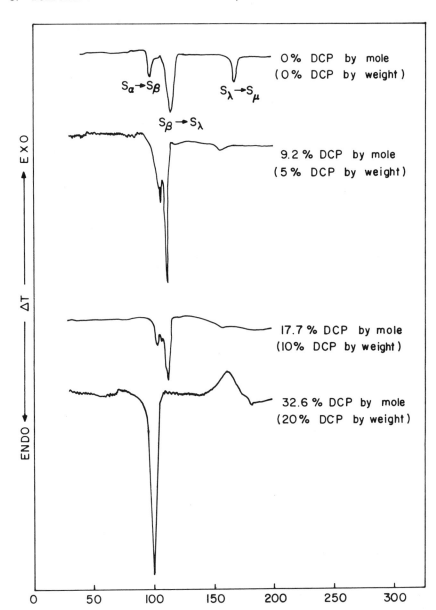

Figure 1. DSC thermograms of sulfur–DCP mixtures with varying compositions

are summarized below. The ampoules have been initially shaken to mix and dissolve the liquid sulfur with the olefin. Only the norbornene unsaturation is responsible for sulfur-olefin copolymerization, although DCP appears to copolymerize more efficiently than norbornene.

$S_8 +$ [structure] $\xrightarrow[140°C]{3\ hr}$ Highly viscous liquid at 140°C; on cooling gives a dark brown gum-like solid, soluble in CS_2

$S_8 +$ [structure] $\xrightarrow[140°C]{6\ hr}$ Very high viscous liquid at 140°C; on cooling gives a dark brown brittle solid, slowly soluble in CS_2. (With commercial DCP and laboratory-grade sulfur, the product is insoluble in CS_2.)

$S_8 +$ [structure] $\xrightarrow[140°C]{3-6\ hr}$ Viscous liquid at 140°C; on cooling gives a yellowish-brown gum-like solid, soluble in CS_2. (The color darkens with increasing time.)

$S_8 +$ [structure] $\xrightarrow[140°C]{6\ hr}$ Mobile liquid at 140°C; on cooling sulfur precipitates out, easily soluble in CS_2

The 60-MHz NMR spectra of norbornene, *endo*-DCP, and *exo*-DCP in CS_2 using tetramethyl silane as the internal standard are shown in Figures 2, 3, and 4, respectively. The norbornene unsaturations in *endo*- and *exo*-DCP are more downfield than the respective cyclopentene unsaturations (9). Furthermore, the cyclopentene unsaturation in *endo*-DCP gives a singlet and in *exo*-DCP a multiplet, thus rendering it possible to distinguish easily between the two isomers (9). In norbornane, **V**, the C-2-exo and C-2-endo protons have chemical shifts of 1.49 δ and 1.18 δ, respectively (10), and thus can be distinguished.

The cis-endo protons in *exo*-3,4,5-trithiatricyclo [5.2.1.0^{2,6}] decane, **VI**, have a chemical shift of 3.60 δ (11). The chemical shift of the cis-

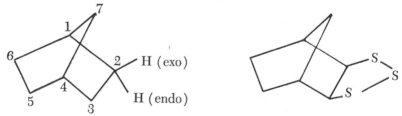

V   VI

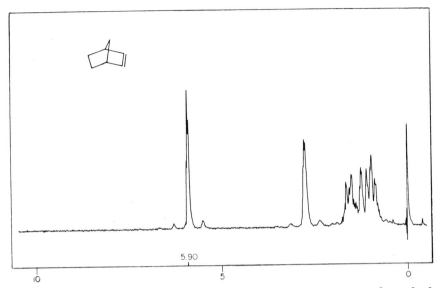

Figure 2.   NMR spectrum of norbornene in CS₂ using TMS as internal standard

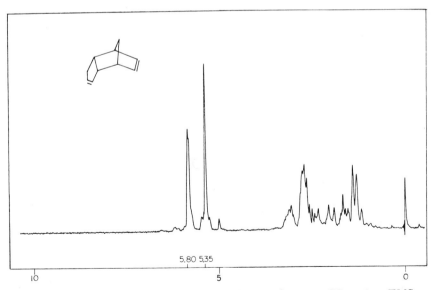

Figure 3.   NMR spectrum of endo-*dicyclopentadiene in CS₂ using TMS as internal standard*

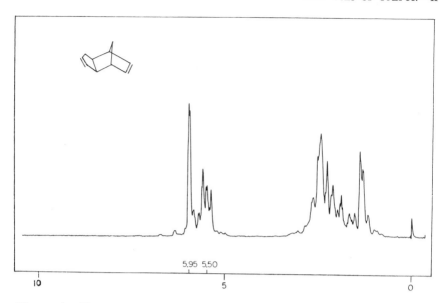

5.95 5.50

10   5   0

Figure 4. NMR spectrum of exo-dicyclopentadiene in CS₂ using TMS as internal standard

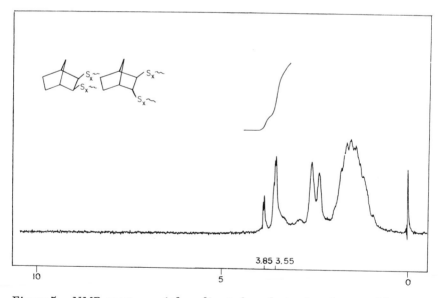

3.85 3.55

10   5   0

Figure 5. NMR spectrum of the sulfurated products of norbornene (almost an equimolar mixture of cis-exo and trans addition products as shown) in CS₂ using TMS as internal standard

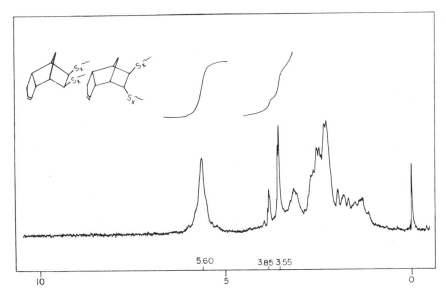

Figure 6. NMR spectrum of the sulfurated products of endo-*dicyclopentadiene (almost an equimolar mixture of cis-exo and trans addition products as shown) in CS₂ using TMS as internal standard*

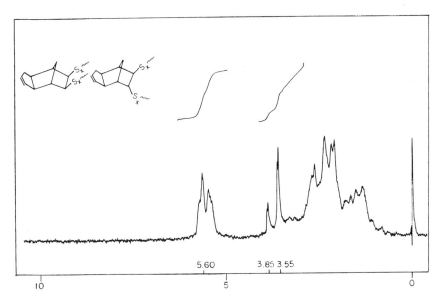

Figure 7. NMR spectrum of the sulfurated products of exo-*dicyclopentadiene (almost an equimolar mixture of cis-exo and trans addition products as shown) in CS₂ using TMS as internal standard*

exo protons of the hypothetical endo isomer of this compound is calculated to be at 3.91 δ. The NMR spectra of the polysulfide products formed
after the 3-hr sulfur–olefin reactions at 140°C with norbornene, *endo*-
DCP, and *exo*-DCP are shown in Figures 5, 6, and 7, respectively, and
indicate that the norbornene unsaturation has reacted completely. The
singlet and multiplet cyclopentene unsaturations are observed around
5.60 δ in Figures 6 and 7, respectively, showing that the cyclopentene
unsaturation does not react with sulfur, nor is there any Meerwein–Wagner type of rearrangement (*12, 13*). The relatively broad nature of the
signals is caused by the polymeric structure of the polysulfides. In elec-
trophilic additions of norbornene, there are one or more of three possible
types of major addition products: 2,3-*cis-exo*, 2,3-*trans*, and 2-*exo*-7-syn
(Meerwein–Wagner rearranged product) (*14*). In Figures 5, 6, and 7
there are two signals of interest; one at 3.55 δ and the other at 3.85 δ.
Considering the examples of 3,4,5-trithiatricyclo [5.2.1.0$^{2, 6}$] decane mentioned earlier, these chemical shifts are attributed to a mixture of cis-exo
and trans sulfurated products: **VII a, VII b, VIII a, VIII b, IX a,** and
IX b. The relative intensities of these signals show that the cis-exo and
trans structures are about equimolar in composition in all three cases,
and therefore it appears to be an equilibrium composition.

VIIa   VIIIa   IXa

VIIb   VIIIb   IXb

These polymeric polysulfide products alternate in structure, as shown,
and are not random because the relative intensities for the olefinic protons
at 5.60 δ and the endo/exo protons at 3.55 δ/3.85 δ combined together
are almost equal, as seen in Figures 6 and 7. It appears to be a "step-

growth copolymerization," because the norbornene unsaturation in DCP disappears in the early stage of the reaction as evidenced by NMR analysis while the viscosity increases even afterwards because of an increased molecular weight as shown by the decreasing ease of solubility in CS_2 with increasing reaction time (15).

The formation of viscous liquids containing polymeric polysulfides from the reaction of sulfur with a number of olefinic hydrocarbons has been reported (16). Blight, Currell, and their colleagues have analytically characterized the sulfur–DCP system and have shown that the increase in viscosity for the sulfur–DCP solutions at 140°C is caused by the formation of high-molecular-weight polysulfides (17). Also as reaction time proceeds, the molecular weight of the polysulfides increases (17).

Currell and his students from Thames Polytechnic, London have reported in a private communication that the 2,3-cis-endo protons in

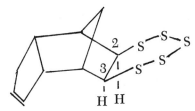

which is extracted by hot chloroform from the sulfurated products of DCP and is separated by preparative GPC, appears to give the NMR signal at 3.8 δ. In that case all the sulfurated products from the bicyclo-[2.2.1]heptene unsaturation would be only cis-exo with no trans. Further work by these authors is in progress.

Melt Viscosity of Sulfur–Dicyclopentadiene Solutions

The behavior of melt viscosity of sulfur–dicyclopentadiene solutions is of obvious interest from the point of sprayable coatings. The melt viscosity behavior has been reported recently, but only qualitatively and over a narrow range of compositions (18). The viscosity of sulfur measured by the capillary method by Bacon and Fanelli (19, 20) is considered to be the best (21). Recently, however, the viscosity of sulfur has been measured by an apparatus containing an electric motor and a rotating cylinder (22). Viscosity of the sulfur–DCP solutions are measured here with the help of a Brookfield synchro-lectric viscometer, which is of the later kind. Viscosity measurements have been carried out to follow the copolymerization reaction and to analyze the viscosity behavior.

The exponential behavior of viscosity as a function of time for varying compositions is shown in Figures 8 and 9. The pronounced effect of temperature on viscosity is shown in Figure 10. Pryor (23) pointed out

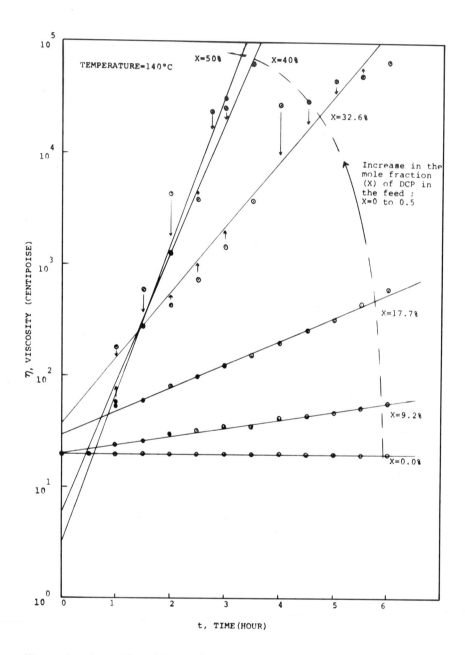

Figure 8.   A semi-logarithmic plot of viscosity vs. time for the sulfur–DCP solutions at 140°C for the composition of DCP mole fraction, X = 0–50%

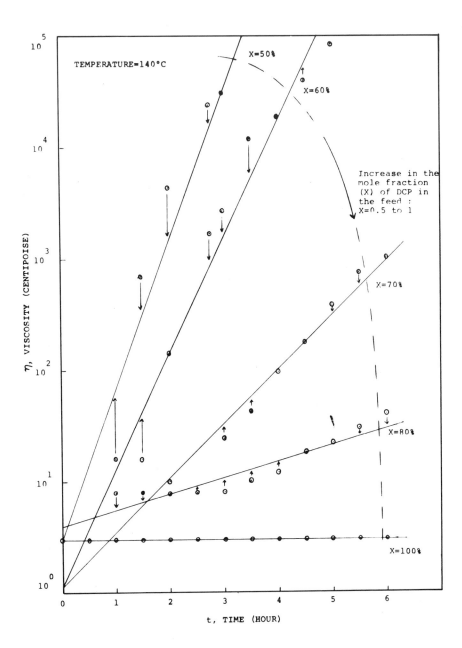

Figure 9. A semi-logarithmic plot of viscosity vs. time for the sulfur–DCP solutions at 140°C for the composition of DCP mole fraction, X = 50–100%

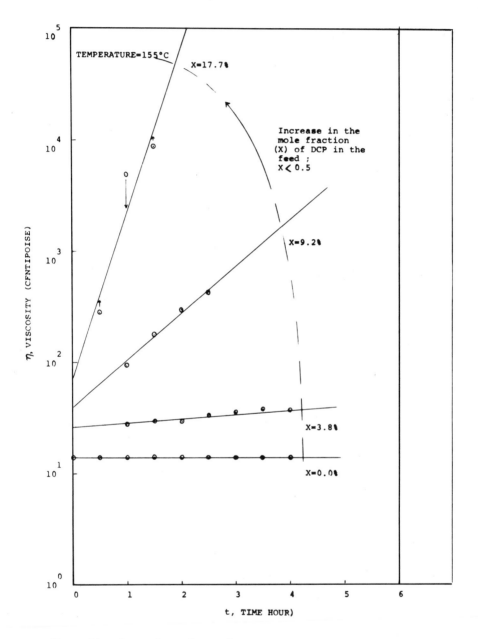

Figure 10. A semi-logarithmic plot of viscosity vs. time for the sulfur–DCP solutions at 155°C for the compositions of DCP mole fraction,
$$X = 0\text{–}17.7\%$$

that in sulfur–olefin reactions, only the C–S bond is formed, and no new C–C bond is formed. Therefore in such a copolymerization reaction, one expects to get a strictly alternating structure. The maximum rate for the logarithmic increase in viscosity for an equimolar feed of S_8 and DCP (Figure 11) indicates that the copolymer is probably alternating in structure, which is consistent with the NMR results. Also, a "step growth copolymerization" mechanism perhaps supports such behavior of viscosity. The linear relationship in Figure 12 shows that the rate of increase

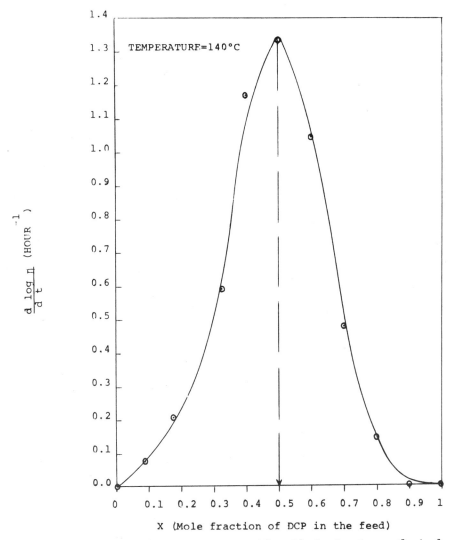

Figure 11. A plot of the rate of increase of logarithmic viscosity vs. the feed composition in mole fractions of DCP at 140°C

44

of the logarithmic viscosity is proportional to X^b . . . except around $X =$ 0.5, where X is the mole fraction of DCP in the feed and b an empirical constant, positive for X less than 0.5 and negative for X greater than 0.5 at a given temperature. The proportionality does not hold around $X =$

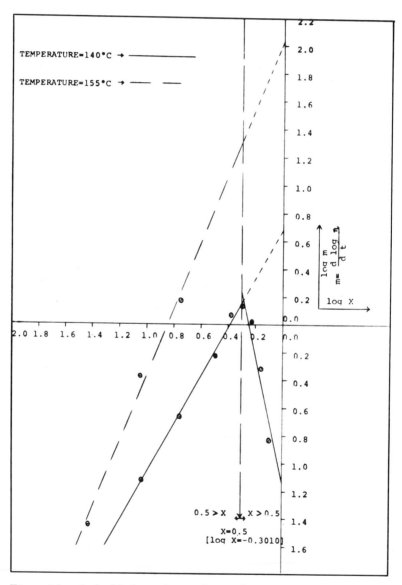

Figure 12. A double-logarithmic plot of the rate of increase of loga-rithmic viscosity vs. the feed composition in mole fractions of DCP; (———) at 140°C and (— — —) at 155°C

0.4–0.6 because of the relatively flat nature of the curve at $X = 0.4$–0.6 in Figure 11.

A general exponential equation (*see* Appendix for detailed discussions) of the form $\eta = \eta_0 \exp(aX^b t)$, at constant temperature, has been developed to predict the viscosity (η, CP) as a function of feed composition (X, mol fraction of DCP) and time (t, hr); η_0, a, and b are empirical constants; η_0 is the viscosity at $t = 0$, a is in units of reciprocal time, and b is dimensionless. The assumptions made here are:

(1) Newtonian behavior of the sulfur–DCP solutions

(2) $(d \log \eta)/dt = a'X^b$, $a' = a/2.30$ at constant temperature, valid for $0 \leq X \leq 0.5$ and $0.5 \leq X < 1$ with two sets of values for a and b.

(3) η_0 depends on temperature only and not on composition to a first approximation because it is an average value determined individually for $X < 0.5$ and $X > 0.5$.

Experimental results appear to support these assumptions. The final viscosity equations with the range of compositions and temperatures at which they are valid are shown below. The parameters η_0, a, and b have been determined by the method of least square approximations using a minicomputer based on experimental data in Figures 8, 9, 10, and 12.

$$\eta^{140^\circ C}_{X < 0.5} = 19.46 \exp(11.33X^{1.78} t), \, [X = 0\text{–}0.4]$$

$$\eta^{140^\circ C}_{X > 0.5} = 2.50 \exp(0.16X^{-4.64}t), \, [X = 0.6\text{–}1]$$

$$\eta^{155^\circ C}_{X < 0.5} = 38.06 \exp(251.66X^{2.42}t), \, [X = 0\text{–}0.2]$$

Surface Tension of Sulfur–Dicyclopentadiene Solutions

The behavior of surface tension along with viscosity is of practical importance for sprayable coatings and also is expected to give some understanding of the polymerization reaction. The bubble pressure method is an experimentally simple method and a good procedure for surface tension measurements of viscous liquids (*24*). Fanelli's (*25*) measurements on the surface tension of sulfur over its entire liquid range using the Sugden's double capillary modification of the maximum bubble pressure method are considered to be the most accurate (*21*). Recently the surface tensions of sulfur were determined in an atmosphere of sulfur vapor in the temperature range 120°–435°C by using the large drop method (*26*). Surface tension measurements have been done in our studies on sulfur–DCP solutions using the former method.

The surface tensions of sulfur and DCP at 140°C are 69.0 and 21.9 dyn cm⁻¹, respectively. The surface tensions of sulfur–DCP solutions at

140°C with two feed compositions are shown in Tables I and II. Surface tension has been measured as a function of time, and the viscosity of the solutions are shown along with surface tension. The data clearly show that as the viscosity increases with time, surface tension increases, and the higher the rate of increase of viscosity, the higher the rate of increase of surface tension. It has been shown for silicone polymers that as the viscosity increases from an increase in molecular weight, the surface tension increases (27). A "step growth copolymerization mechanism," as mentioned earlier for the sulfur–DCP solutions, will have an increase of molecular weight with time, and the surface tension behavior appears to support this mechanism.

Table I.   Sulfur–5 Wt % DCP (or 9.2 Mol %) at 140°C

Time (hr)	Viscosity (cP)	Surface Tension (dyn cm⁻¹)
0	16	62.7
1	24	62.7
2	30	65.9
3	35	70.6
4	42	73.7
5	47	76.8

Table II.   Sulfur–20 Wt % DCP (or 32.6 Mol %) at 140°C

Time (hr)	Viscosity (cP)	Surface Tension (dyn cm⁻¹)
0	26	50.2
1	184	75.3
2	430	84.7
3	1430	not possible to measure

Sulfur–Dicyclopentadiene Copolymeric Material as Compared with Plastic Sulfur

When molten sulfur is heated above 159°C, preferably to 200°–250°C, and then rapidly quenched to about −20°C, a translucent elastic dark-brown material (plastic sulfur) is obtained (28). Plastic sulfur, which is a mixture of amorphous S_8 rings and amorphous polymeric sulfur, is thermodynamically unstable (28). It undergoes embrittlement, especially above −10°C, because of the rapid crystallization of octameric sulfur to orthorhombic sulfur (28). Also polymeric sulfur depolymerizes and crystallizes to orthorhombic sulfur slowly at ambient temperature and rapidly above 90°C (28).

Materails made from sulfur with 20 wt % (or 25% w/w) DCP at 140°C for 3 hr contained 45.6% polysulfides and 54.5% unreacted sulfur,

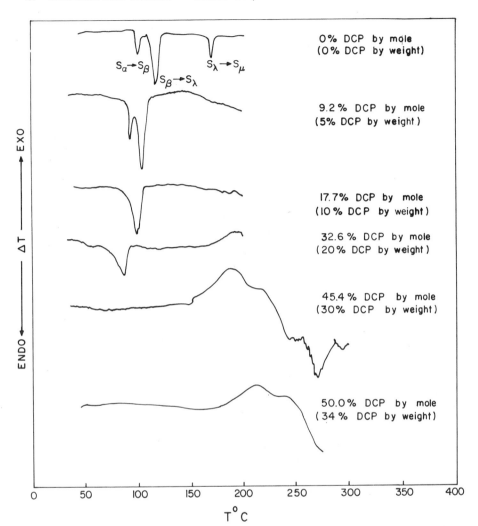

Figure 13.   DSC thermograms of the sulfur–DCP reaction product with varying feed compositions, after 6 hr of reaction at 140°C and after 6 mo of storage under ambient conditions

out of which 30.5% crystallized out as monoclinic, and the rest remained at an amorphous state after 18 months storage (29). The DSC thermograms of the sulfur–DCP products made with 6 hr of reaction time at 140°C and after six months storage under ambient conditions (Figure 13) show that orthorhombic/monoclinic sulfur crystallized out in decreasing amounts with increasing amounts of DCP, and no orthorhombic or monoclinic sulfur crystallized out from feeds of about equimolar composition (or 34 wt % DCP). A correlation is drawn between the feed composition,

Table III.   Relationship of the Sulfur–DCP Composition, Melt
Viscosity at 140°C, and DSC Thermal Transition of
the Orthorhombic/Monoclinic Sulfur

S_8–DCP Feed Composition		Melt Viscosity at 140°C and at the 6th Hr in cP	DSC-Endotherm of S_α/S_β Transition of the Product after 6 Mo Storage
% DCP by Weight	% DCP by Mole		
0	0	20	present
5	9.2	58	present
10	17.7	612	present
20	32.6	68,000	present
30	45.4	> 100,000	absent
34	50.0	>> 100,000	absent

melt viscosity at the end of the reaction at 140°C, and the crystallization
of sulfur, as shown in Table III. The high viscosity for the equimolar feed
composition shows high conversion of sulfur to sulfur–DCP copolymer
from which no crystallization of sulfur takes place. However the sulfur–
DCP copolymeric materials thus obtained are brittle. It appears that if
a proper plasticizer can be developed, DCP can be used successfully to
stabilize and plasticize plastic sulfur by copolymerization with sulfur.

Dodecylpolysulfide as a Viscosity Suppressor and as a Plasticizer
for the Sulfur–Dicyclopentadiene System

A plasticizer is a compatible low-melting-point solid or a high-boiling
point liquid which when added to a rigid and brittle material in relatively

Table IV.   Effect of Dodecylpolysulfide

Composition of S_8:DCP:DDPS		Melt Viscosity Behavior at 140°C		
		Viscosity, η (cP), at Time, t (hr)		
By Mole	By Weight	t = 0	t = 3	t = 6
1:1:0.0	S → 66.0% DCP → 34.0% DDPS → 0.0%	8	3.1×10^4	>> 10^5 too high to measure
1:1:0.1	S → 58.9% DCP → 30.4% DDPS → 10.7%	14	6.0×10^2	4.2×10^3
1:1:0.2	S → 53.2% DCP → 27.4% DDPS → 19.3%	14	3.5×10	7.0×10

small amounts imparts flexibility (30). The fundamental principles involved in understanding the role of additives in plastics have been considered by Mascia (31). Alkyl disulfides, e.g., tertiary butyl disulfide or tertiary dodecyl disulfide are insoluble in the liquid sulfur–DCP system, so they cannot possibly be plasticizers. Alkyl polysulfides, e.g., tertitary butyl polysulfide of sulfur rank 4–5, are soluble in the liquid sulfur–DCP system at 140°C. However, this material possesses a very disagreeable odor which is probably caused by its relatively low boiling point and high vapor pressure, and therefore it cannot be used as a plasticizer from a practical point of view.

Dodecyl polysulfide, DDPS [$CH_3(CH_2)_{11}$ $S_n(CH_2)_{11}$ CH_3, where the average value of $n = 4$] has been synthesized to investigate its properties as a potential plasticizer. It possesses almost no odor because of its low vapor pressure reflected in its high boiling point (greater than 300°C), above which decomposition occurs.

DDPS has shown two-fold promise with the sulfur–DCP system in our preliminary in situ studies:

(1) It acts as a diluent that effectively reduces the viscosity and the rate of viscosity increase of the sulfur–DCP solutions.

(2) It acts as a plasticizer to give a flexible solid material at ambient temperature. The flexibility will, of course, depend on the amount of DDPS used as well as on the reaction time and temperature for a given sulfur–DCP composition.

Some of the experimentally determined data are summarized in Table IV. An equimolar composition of S_8:DCP was chosen there because

on the Sulfur–DCP System

Melt Viscosity Behavior at 140°C	
Rate of Increase of Logarithmic Viscosity $d \log_{10}\eta/dt$ (hr^{-1})	*Properties of the Materials at Ambient Temperatures Made from 6 Hr Reaction at 140°C*
1.34	dark-brown hard brittle solid
0.39	dark-brown soft flexible solid
0.12	dark-brown very high-viscous liquid

the rate of increase of viscosity is greatest for this composition, and ortho-rhombic or monoclinic sulfur has no tendency to crystallize out from this composition.

Experimental

Chemicals. SULFUR. For viscosity and surface tension experiments, laboratory-grade sublimed powdered flowers of sulfur from Fisher Scientific Co. are used without further purification. For synthetic and characterization purposes, 99.9995% sulfur from PCR Inc. is used.

DICYCLOPENTADIENE. For visocity and surface tension experiments DCP from Exxon Chemical Co. is used without further purification. This is around 98% pure, with the remainder primarily acyclic dienes. The isomer composition is 98.5% endo and 1.5% exo.

endo-DICYCLOPENTADIENE. DCP from Exxon Chemical Co. is distilled twice using a Vigreaux column under reduced pressure; bp, 93°C at 5.8 cm Hg, 170°C at 76.0 cm Hg; mp, 33.6°C.

exo-DICYCLOPENTADIENE. First 2,3-dihydro-2-bromo-exo-DCP is synthesized from endo-DCP (32), which is then dehydrobrominated using potassium tertiary butoxide/dimethyl sulfoxide to give exo-DCP (13), bp 48°C at 1.0 cm Hg and liquid at room temperature.

2,3-DIHYDRO-endo-DCP. It is synthesized by partial hydrogenation of endo-DCP using 10% palladium–charcoal catalyst at 3 atm and at room temperature (33). It has a mp of 52°C, and the reported bp is 181°C at 76.8 cm Hg (34). The NMR spectrum had one olefinic singlet at 5.5 δ (9).

2,3-DIHYDRO-exo-DCP. It is synthesized in the same method as above but starting with exo-DCP. It has a bp of 54°C at 1.0 cm Hg, and it is a liquid at room temperature. It shows one olefinic multiplet at 5.5 δ in the NMR spectrum (9).

NORBORNENE. 99% Norbornene from Aldrich Chemical Co. is purified by sublimation. (mp, 46°C; bp, 96°C)

DODECYL POLYSULFIDE. Distillation-purified dodecyl mercaptan from Eastman (0.626 mol) and sulfur (0.919 g-atom) are placed in a three-necked flask fitted with a mechanical stirrer and a condenser. Oxidation of mercaptan is catalyzed by 0.2 ml diethyl amine at 50°C. Although the reaction is practically complete in about 6 hr, it was held for 22 hr to ensure complete conversion of mercaptan to polysulfide. Then it was heated at 75°C for 2 hr without a condenser to remove as much Et_2NH (bp = 55°C) as possible. DDPS [$CH_3 (CH_2)_{11} S_n (CH_2)_{11} CH_3$ with an average value of $n = 4$], is a transparent yellow liquid with a density of 0.965 g ml^{-1} at 20°C and decomposes on boiling. Molecular weight has been determined to be 470 (theoretical for $n = 4$ is 466.5). Elemental analysis shows 62.42% C, 10.75% H, and 26.42% S; calc. 61.79% C, 10.72% H, and 27.49% S. (Elemental analysis and molecular weight determinations were carried out by Galbraith Laboratories, Inc.)

Procedures. In viscosity/surface tension measurements, DCP is added slowly while stirring to liquid sulfur in a tall form beaker/bubbler held in a paraffin oil constant-temperature bath.

In synthesis–characterization experiments, sulfur–olefin samples are taken in 2-ml glass ampoules, frozen with dry ice–acetone and sealed under vacuum. Reactions are run in an oven at 140°C, and the ampoules are shaken in the beginning to dissolve the liquid sulfur with the olefins.

For thermal analysis a nitrogen atmosphere and a heating rate of 10°C/min are used. For the NMR analysis, CS_2 is used as the solvent with tetramethyl silane as the internal standard.

Equipment. A Brookfield synchro-lectric viscometer, serial no. 758, is used to measure viscosity in the range of 0–100,000 cP. Sugden's double capillary modification of the maximum bubble pressure method is used to determine surface tensions. The apparatus is calibrated with benzene and is checked by determining the surface tension of chloroform at 25°C, which is found to be 23.5 dyn cm^{-1} (26.5 dyn cm^{-1}) (35).

A DuPont 900 thermal analyzer equipped wtih a DuPont DSC cell is used for thermal analysis. A Varion EM-360 60-MHz NMR spectrometer is used for NMR structure analysis.

Appendix

Melt Viscosity Equation of Sulfur–DCP Solutions. An assumption is made that the sulfur–DCP solution is a Newtonian fluid, i.e., the viscosity measured by the Brookfield viscometer is independent of the spindle speed, which is related to shear rate. The linear plots of log(viscosity) vs. time as in Figures 8, 9, and 10 give the following equation for a given sulfur–DCP composition at a given temperature:

$$\log_{10} \eta = mt + \log_{10} \eta_0' \tag{1}$$

where η is the viscosity (cP) in time t (hr), η_0' the viscosity at $t = 0$, and m the slope, $d \log_{10} \eta/dt$.

An average value of η_0', s of the different compositions (for mole fraction of DCP < 0.5 and > 0.5) at a given temperature can be considered to be independent of composition for a first-order approximation and is denoted by η_0. Substituting η_0, Equation 1 gives:

$$\log_{10} \eta = mt + \log_{10} \eta_0 \tag{2}$$

or

$$\eta = \eta_0 \exp(2.30mt) \tag{3}$$

Let's assume that m is related to the concentration of DCP in the feed in mole fraction X at constant temperature by Equation 4:

$$m = a'X^b \tag{4}$$

where a' and b are emperical constants dependent on temperature only, b is a dimensionless number ($+$ ve for $0 \leq X \leq 0.5$ and $-$ ve for $0.5 \leq X < 1$), and a' is in the units of time^{-1}.

Equation 4 can be written as:

$$\log_{10} m = b \log_{10}X + \log a' \qquad (5)$$

The linear plots of $\log m$ vs. $\log X$ at 140° and 155°C for $X < 0.5$ and $X > 0.5$ in Figure 12 show the validity of the above assumption, except around $X = 0.4–0.6$. Equation 4 is not expected to hold in that region because of the relatively flat nature of the curve around $X = 0.4–0.6$ in Figure 11.

Therefore the viscosity of sulfur–DCP solutions at a given temperature is a function of composition and time only and can be represented by:

$$\eta = \eta_0 \exp(2.30\, a'X^b t) \qquad (6)$$

Putting

$$a = 2.30\, a' \qquad (7)$$

the general viscosity equation is given by: $\eta = \eta_0 \exp(aX^b t)$, at constant temperature, which is valid as a first-order approximation that η_0 is a function of temperature only and not of composition.

Acknowledgment

The authors wish to thank the Sulphur Institute for partial financial support for this research and also H. L. Fike, Director, Industrial Research of the Sulphur Institute, Washington, D.C., for his interest and many helpful discussions.

Literature Cited

1. Fike, H. L., "Sulfur Research Trends," ADV. CHEM. SER. (1972) 110, Chap. 16.
2. "Statistical Supplement," No. 13, pp. 1–4, British Sulphur Corporation Ltd., London, 1976.
3. Dale, J. M., "Sulfur-Fibre Coatings," *Sulphur Inst. J.* (Fall 1965) 1 (1), 11–13.
4. Dale, J. M., Ludwig, A. C., "Sulfur Coatings for Mine Support," Final Report on Project 11-3124, Southwest Research Institute, San Antonio, TX 78284, 1972.
5. Sullivan, T. A., McBee, W. C., "Sulfur Utilization in Pollution Abatement," *Proc. Miner. Waste Util. Symp., 4th* (May 7–8, 1974).
6. West, J. R., "New Uses of Sulfur," ADV. CHEM. SER. (1975) 140, Chaps. 2, 3, 4, 10.
7. Cesca, S., *Macromol. Rev.* (1975) 10, 123–131.

8. Bateman, L., Moore, C. G., in "Organic Sulfur Compounds," N. Kharasch, Ed., Vol. 1, Chap. 2, Pergamon, New York, 1961.
9. Cesca, S., Santostase, M. L., Marconi, W., Palladino, N., *Ann. Chim.* (Rome) (1965) **55**, 704–729.
10. Tori, K., Aono, K., Hata, Y., Muneyuki, R., Tsuji, T., Tanida, H., *Tetrahedron Lett.* (1966) (1), 9–17.
11. Shields, T. C., Kurtz, A. N., *J. Am. Chem. Soc.* (1969) **91** (19), 5415–16.
12. Kennedy, J. P., "Cationic Polymerization of Olefins: A Critical Inventory," pp. 220, 226–228, Wiley Interscience, New York, 1975.
13. Corner, T., Foster, R. G., Hepworth, P., *Polymer* (1969) **10** (5), 393–397.
14. Cristol, J., Arganbright, R. P., Brindell, G. D., Heitz, R. M., *Am. Chem. Soc.* (1957) **79**, 6035–6038.
15. Billmeyer, F. W., Jr., "Textbook of Polymer Science," 2nd ed., p. 256, John Wiley & Sons, New York, 1971.
16. Williams, A. J., Ph.D. Thesis, "Interaction of Sulphur with Olefinic Hydrocarbons," submitted in the University of London, London, England, 1973 (B. R. Currell, research director).
17. Blight, L., Currell, B. R., Nash, B. J., Scott, R. A. M., Stillo, C., ADV. CHEM. SER. (1978) **165**, 13.
18. Diehl, L., "Dicyclopentadiene (DCP)—Modified Sulfur and Its Use as a Binder, Quoting Sulfur Concrete as an Example," Madrid (May 18, 1976). Copy in English available at the Sulphur Institute, Washington, D. C.
19. Bacon, R. F., Fanelli, R., *J. Am. Chem. Soc.* (1943) **65**, 639–648.
20. Fanelli, R., *J. Am. Chem. Soc.* (1945) **67**, 1832–1834.
21. W. N. Tuller, Ed., "The Sulphur Data Book," pp. 19–23, McGraw-Hill, New York, 1954.
22. Burba, A. A., Kerdinskii, M. E., Ignachkov, V. I., *Zavod. Lab.* (1974) **40** (1) 62–63 (Russ.) [*Chem. Abstr.* (June, 1974) **80**, (11) 122730 k.]
23. Pryor, W. A., "Mechanisms of Sulfur Reactions," Chap. 5, McGraw Hill, New York, 1962.
24. Weissberger, A., Rossiter, B. W., Eds., "Physical Methods of Chemistry," Vol. I, Part V, Chap. IX, Wiley Interscience, New York, 1971.
25. Fanelli, R., *J. Am. Chem. Soc.* (1950) **72**, 4016–4018.
26. Dokhov, M. P., Zadumkin, S. N., Karashaev, A. A., Unezhev, B. Kh., *Zh, Fiz. Khim.* (1976) **50** (7) 1801–1803 (Russ.) [*Chem. Abstr.* (1976) **85**, 149387 u.]
27. Allan, G. G., Neogi, A. N., *J. Appl. Polym. Sci.* (1970) **14**, 999–10005.
28. Tobolsky, A. V., MacKnight, W. J., "Polymeric Sulfur and Related Polymers," Chaps. 1, 5, Interscience, 1965.
29. Currell, B. R., Williams, A. J., Mooney, A. J., Nash, B. J., in "New Uses of Sulfur," ADV. CHEM. SER. (1975) **140**, Chap. 1.
30. Meyer, B., Ed., "Elemental Sulfur," Chap. 8, Interscience, New York, 1965.
31. Mascia, L., "The Role of Additives in Plastics," John Wiley & Sons, New York, 1974.
32. Bruson, H. R., Riener, T. W., *J. Am. Chem. Soc.* (1945) **67**, 1178–1180.
33. Foster, R. G., McIvor, M. C., *J. Chem. Soc. (B)* (1969) 188–192.
34. Sesca, S., Santostasi, M. L., Marconi, W., Greco, M., *Ann. Chim. (Rome)* (1965) **55**, 682–703.
35. Fanelli, R., *J. Am. Chem. Soc.* (1948) **70**, 1792–1793.

RECEIVED April 22, 1977.

4

A New Approach to Sulfur Concrete

R. GREGOR and A. HACKL

Institut fur Verfahrenstechnik, Technical University of Vienna,
A-1060 Vienna, Austria

Sulfur concrete may be attained with a compressive strength of approximately 1200 kp/cm² by using 3 wt % DCP-modified sulfur as binder and defined grain size distributions for aggregates. Microscopic examinations of thin polishes show the importance of optimally composed aggregates to the strength of sulfur concretes. Higher DCP concentration results in lower compressive strength but higher flexural strength. With 10 wt % DCP-modified sulfur, one can achieve a flexural strength of over 200 kp/cm². Examinations by ESCA and SEM showed that sulfur does not bind preferentially to certain mineral parts.

Product Definition

In the melted state elementary sulfur is an excellent binder for aggregates such as sand, gravel, crushed stone, and similar materials. When a hot sulfur aggregate mixture is left to set, a material of concrete-like hardness is obtained. This property leads to the term sulfur concrete or sulfur mortar, which is wrong strictly speaking, since the word concrete is used to refer to a product in which aggregates are bound with hydraulic products such as cement or with limestone. However, since polymer-bound aggregates recently developed to industrial maturity have been called polymer or synthetic resin concrete, the term sulfur concrete is maintained in this chapter.

As early as 1859 Wright mentioned the binding property of elementary sulfur (*1*), but up to the 1930s mixtures of sulfur and stone powders were used exclusively to grout foundation screws. The first references in the literature to the chemical and physical properties of sulfur mortar are found in the works by Duecker and Payne (*2, 3, 4, 5*). The sulfur mortar described by them contained about 60 wt % sulfur and therefore only a small amount of aggregate. It was used as a sealing mass for the construction of acid tanks and pickling baths which were lined with acid-resistant bricks.

In the late 1960s experiments were made again to produce sulfur concrete. Now, the emphasis was on developing a construction material (6, 7). The first products had a sulfur content of about 30 wt % and had compressive and flexural strengths in the order of 250 kp/cm² and 35 kp/cm², respectively. These first experiments were followed by more research (8, 9, 10, 11).

However, when starting our work in 1975, a series of questions regarding this construction material were not yet clarified to us. We wanted to answer these questions by our work.

Development of High-Strength Sulfur Concretes

One of the most conspicuous advantages of sulfur concrete compared with cement concrete, besides its chemical resistance, is its strength development (Figure 1). Even with very strong and thus early-setting cements, cement concrete requires about 28 days for hydration in order

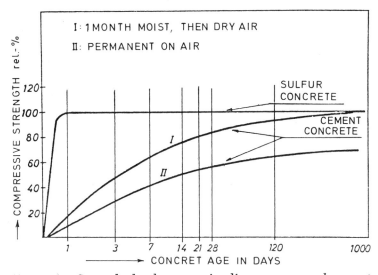

Figure 1.   Strength development of sulfur concrete and cement concrete as function of concrete age and moisture supply

to achieve about 90% of its final strength if certain moisture and temperature conditions are observed. Sulfur concrete, on the other hand, reaches this final strength after only a few hours, and moisture and temperature do not influence the development of strength. Sulfur concrete can be produced therefore at temperatures below freezing.

Unmodified Sulfur Concrete

In the first part of this chapter we examine the influence of grain distribution, grain form, and basicity of aggregates on the strength of sulfur concrete.

Materials Used. The sulfur used was obtained from the refinery and was 99.9% pure. As for the grain form, gravel as spherical material and crushed stone (granulit, basalt) as sharp grain material were used as aggregates. In mineral composition, granulit is an acidic rock, but basalt, on the other hand, consists mainly of basic mineral components.

The aggregates used were screened in order to obtain defined grain size distributions (Tables I, II, and II). The largest grain was limited

Table I. Physical Properties of Aggregates Used

Sample No.	Aggregate Type	Specific Gravity (kg/dm^3)	Compressive Strength (kp/cm^2)
10 x	gravel	2.64	2,200
20 x	granulit	2.65	2,400
30 x	basalt	2.90	3,100

Table II. Quantitative Analysis of Aggregates Used

Element (%)	Gravel	Granulit	Basalt
SiO_2	33.3	70.7	42.7
Al_2O_3	—	15.5	16.6
Fe_2O_3	—	3.4	10.9
K_2O	—	3.8	2.5
Na_2O	—	4.9	5.0
CaO		—	—
MgO	rest	—	—
TiO_2	—	—	—
Loss at red heat	26.6	—	—

Table III. Sieve Analysis of Aggregate Used[a]

Sieve Line	Square Sieve Openings (mm)						
	0.9	0.2	0.5	1	2	4	8
1	—	7.25	18.50	30.00	46.50	68.50	100
2	3.75	3.50	18.50	30.00	46.50	68.50	100
3	9.00	14.50	25.70	37.80	54.20	75.79	100
4	10.61	15.81	25.00	35.36	50.00	70.71	100
5	16.62	22.87	32.99	43.54	57.43	75.79	100
6	26.02	33.07	43.53	53.59	65.98	81.23	100

[a] Standard sieves according to DIN 4188 Blatt 1; DIN 4187 Blatt 2. Cumulative percentage passing each sieve (wt %).

to 8 mm. For testing purposes cubes with edge lengths of 7.1 cm and 10 cm, respectively, and prisms 4 × 4 × 16 cm were produced by casting the hot sulfur aggregate mixture and using vibration compaction. Any protruding sections were cut off with a diamond saw after cooling.

Assessment of the Binder Demand. The quantity needed was calculated by ascertaining experimentally the vibration density of the aggregates used in any given case. From the vibration density given on the vibration table, the void space of the grain heap was calculated using the following equation:

$$PV = 100 - \frac{S_R}{R_D} \cdot 100$$

where PV = total pore volume of the vibrated mineral mass (vol %), S_R = vibration density of the vibrated mineral mass (g/cm^3), and R_D = raw density of the aggregate (g/cm^3).

Usually 10–20 wt % was added to the pore volume obtained. This portion corresponded to the sulfur necessary to glue the individual grains.

Influence of Aggregate Grain Size Distribution on the Strength and Workability of Sulfur Concrete. In order to manage with low-sulfur contents and to utilize the compressive strength of the sulfur best, it seemed necessary to build up a maximum-bearing mineral structure with little void space. The liquid sulfur was to glue the individual grains to each other and to fill the grain heap pores completely. However, the binder, and thereby the sulfur demand, is not only determined by the void space to be filled but also by the sum of the grain surface. Following an Austrian norm (*12*) for a favorable grain size distribution of the aggregates to produce cement concrete, a defined sieve line area is recommended for aggregates 0/8 (Figure 2).

In the first tests to produce sulfur concrete, the aggregates were built up according to their grain size distribution following the sieve lines 1 and 2 and were therefore within the favorable sieve line range. However, mixtures whose aggregates were composed according to sieve lines 1 and 2 had extremely bad workability, since there was a strong tendency to separate, and no homogeneous products could be produced.

When looking for acceptable grain size distributions, we chose the grain distribution recommended by Fuller (*13*). Fuller's parabola follows the equation:

$$A = 100 \cdot \left(\frac{d}{D}\right)^n \qquad n = 0.5$$

where A = percentage of a grain class d of $0/d$, D = largest grain diameter of the grain mixture, and d = any grain diameter between $0/D$. The

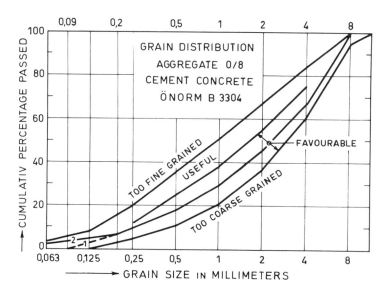

Figure 2. Grain size distribution for aggregates 0/8 to produce cement concrete (ÖNORM B 3304)

Fuller parabola for aggregates 0/8 is represented by sieve line 4 in Figure 3. Sulfur concrete mixtures whose aggregates corresponded to the grain size distribution of the Fuller curve no longer separated and were plastic

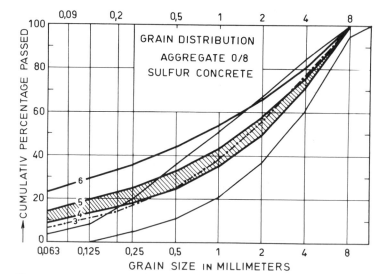

Figure 3. Grain size distribution for aggregate 0/8 to produce sulfur concrete

in their consistency. The values for compressive and flexural strength rose further.

Hummel showed that the Fuller parabola ($n = 0.5$) does not yet furnish the optimal packing densities (*14*). In the case of spherical aggregates $n = 0.4$ ought to be chosen as the greatest density; in the case of crushed aggregates $n = \sim 0.3$ if the density maximum is to be reached for the aggregate (Figure 4).

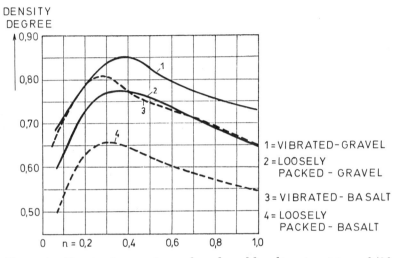

Figure 4.   *Density degree of gravel sands and basalt grain mixtures 0/30 mm grain compositions.* A $= 100 \cdot (d/D)^n$

In tests in which the aggregates were built up according to sieve line 5 (the exponent factor being $n = 0.4$), the void space of the grain heap could be reduced further. With a filler content of 16.62% (thereby it is understood that the fine grain component is smaller than 0.09 mm), the specific surface of the grain mixture was increased further to a considerable degree. Mixtures of this sieve line composition were rather stiff and difficult to compact. Higher compressive and flexural strength values could be reached only by increasing the sulfur and thus the binder content.

Grain compositions built up according to sieve line 6, $n = 0.3$, still have a somewhat lower void space than those built up according to sieve line 5. Because of the high filler content (26%) the quantity of sulfur which is necessary to glue the grains must be increased overproportionally, as compared with the somewhat lower void space, to obtain workable mixtures (Table IV).

The highest strength values were obtained with grain mixtures which were built up according to sieve line 5. The most favorable sieve

Table IV. Data on Individual Samples

Samples No.	Sieve Line	Aggregate Cavity Vibrated (vol %)	Sulfur Content (vol %)	(wt %)	Consistency[a]	Specific Area (m²/kg)[b]
101	1	20.9	25.0	20.7	—[c]	2.24
102	2	21.2	21.2	17.2	K1	3.34
103	2	21.5	23.6	19.5	K3	3.34
104	4	20.4	24.0	19.8	K2	5.93
203	2	25.6	27.0	22.4	K1	3.33
204	3	23.5	28.0	23.3	K3–K2	5.03
205	4	23.4	27.0	22.4	K2	5.91
206	5	21.6	26.0	21.5	K1	8.76
304	3	25.0	28.0	21.9	K2	4.66
305	4	26.0	29.0	22.8	K2–K3	5.48
306	5	20.9	24.0	18.6	K1	8.12
307	5	21.2	26.0	20.3	K1–K2	8.12
308	5	—	28.0	21.9	K2	8.12

[a] K 1 = stiff, K 2 = plastic, K 3 = soft.
[b] Supposition: spherical surface, neglect of microtexture.

$$S = \frac{6}{R_D} \cdot \sum_{i=4}^{n} \frac{ai}{100 \, di}$$

R_D = specific gravity (kg/dm³), ai = portion of the grain class i (wt %), di = ideal middle grain (mm).
[c] Segregation.

line area for the production of sulfur concrete lies within the area limited by the sieve lines 4 and 5. By continuously building up these grain compositions, high packing densities with low grain heap porosity result. Because of the high filler content these sulfur concrete mixtures can be handled extremely well.

The highest compressive and flexural strength values obtained with gravel sulfur concrete were 487 kp/cm² and 95 kp/cm², respectively. The highest compressive and flexural strengths obtained with acidic granulit as aggregate were 645 kp/cm² and 114 kp/cm², respectively. With basic basalt the highest compressive strength was 787 kp/cm², and the highest flexural strength was 112 kp/cm² (Table V).

Favorable grain distributions were found to be very important to the strength of sulfur concrete. Figure 5 shows a thin polish of a sulfur concrete sample whose aggregates are composed according to sieve line 1 and which contains no filler. The compressive strength of this sample was 270 kp/cm², and the flexural strength was 52 kp/cm². Figure 6 shows a sulfur concrete sample where the aggregates are composed according to sieve line 5 and which thus contains 16 wt % filler. In this case the compressive strength was 756 kp/cm², and the flexural strength was 121 kp/cm². In this photograph the grain heap pores are filled completely by the sulfur–filler system.

When comparing the two figures, it can be seen easily how important the filler content of the aggregate as a thickening agent is to reducing the separation tendency and thus to producing homogeneous products. Apart from reducing the inclination to separate, which guarantees better binder distribution during the mixing process, the filler, at the same time, appears to reinforce and support the binder sulfur in the solidified state.

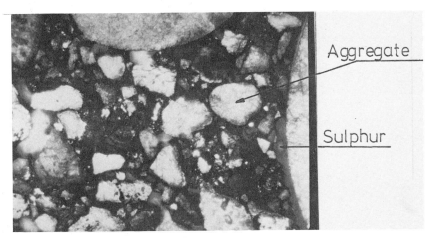

Figure 5.   Nicols II gravel, sieve line 1, sulfur content 20.7 wt % (×25)

Figure 6.   Nicols II basalt; sieve line 5, sulfur content 20.3 wt % (×25)

Table V. Strength Test

Sample No.	Aggregate Type	Density Results (kg/dm^3)	Density Average (kg/dm^3)	Flexural Strength Results (kp/cm^2)	$\overline{X}$ (kp/cm^2)	Sx[d]	Vx[e]
101	kies[a]	2.38		54			
		2.39		50			
		2.41	2.39	52	52	2.0	3.8
102	kies[a]	2.44		95			
		2.43		75			
		2.42	2.43	81	84	10.3	12.3
103	kies[a]	2.44		74.5			
		2.45		74			
		2.42	2.44	80	76	3.3	4.4
104	kies[a]	2.44		93			
		2.45		93			
		2.46	2.45	99	95	3.5	3.6
203	granulit[b]	2.39		68			
		2.40		73			
		2.39	2.39	76	72	4.0	5.6
204	granulit[b]	2.40		97			
		2.41		98			
		2.40	2.40	96	97	1.0	1.0
205	granulit[b]	2.40		106			
		2.41		96			
		2.40	2.40	105	102	5.5	5.4

Results of Sulfur Concrete

	Compressive Strength			
Failure Load		$\overline{\mathrm{X}}$		
(Metric Tons)	*(kp/cm²)*	*(kp/cm²)*	Sx[d]	Vx[e]
7.4	296			
7.00	280			
5.80	232			
7.50	300			
6.80	272			
6.00	240	270	28.4	10.5
13.0	520			
11.6	464			
12.0	480			
10.5	420			
11.6	464			
10.0	400	458	42.9	9.4
12.0	480			
12.0	480			
12.5	500			
12.0	480			
11.5	460			
12.0	480	480	12.6	2.6
12.4	496			
12.5	500			
12.2	488			
11.5	460			
12.25	490			
12.15	486	487	14.1	2.9
11.5	460			
11.5	460			
12.2	488			
11.5	460			
9.8	392			
10	400	443	38.3	8.6
13.7	548			
14.0	560			
14.0	560			
12.7	508			
16.7	668			
14.5	580	571	53.4	9.35
15.6	624			
15.2	608			
16.2	648			
17.3	692			
16.1	644			
16.0	640	643	28.4	4.4

Table V.

Sample No.	Aggregate Type	Density Results (kg/dm^3)	Average (kg/dm^3)	Flexural Strength Results (kp/cm^2)	$\overline{X}$ (kp/cm^2)	Sx[a]	Vx[e]
206	granulit[b]	2.39		112			
		2.37		111			
		2.42	2.39	120	114	4.9	4.3
304	basalt[e]	2.49		92			
		2.48		99			
		2.47	2.48	100.5	97	4.5	4.7
305	basalt[e]	2.53		119			
		2.50		114			
		2.54	2.52	109	114	5.0	4.4
306[f]	basalt[e]	—	—	—	—	—	—
307	basalt[e]	2.52		120			
		2.55		122			
		2.52	2.53	122	121	1.1	0.95
308	basalt[e]	2.53		124			
		2.53		120			
		2.52	2.53	120	121	2.3	1.9

[a] Aggregate, spherical gravel; test age, 4 days.
[b] Aggregate, acidic crushed stone; test age, 4 days.
[e] Aggregate, basic crushed stone; test age, 4 days.

Influence of Grain Form on the Strength of Sulfur Concrete. When the influence of the grain form on the strength of sulfur concrete is examined, complete analogies to cement concrete are found. Because of its spherical form gravel can be compacted more easily and therefore has a lower grain heap porosity. Because of its smooth surface and its spherical form the binder demand is lower.

(Continued)

Compressive Strength

Failure Load		$\overline{X}$		
(Metric Tons)	(kp/cm²)	(kp/cm²)	Sx[d]	Vx[e]
16.7	668			
16.9	676			
15.0	600			
15.0	600			
16.1	644			
17.0	680	645	36.8	5.7
13.5	540			
15.0	600			
11.5	460			
12.5	500			
16.0	640			
12.5	500	540	68.1	12.6
15.5	620			
16.0	640			
17.0	680			
16.5	660			
16.0	640			
16.5	660	650	20.9	3.2
—	—	—		
17.75	710			
19.5	780			
18.90	756			
18.5	740			
19.5	780			
19.25	770	756	27.3	3.0
20	800			
20.25	810			
19.5	780			
19.35	770			
19.5	780			
19.5	780	787	15.1	1.91

[d] Standard deviation.
[e] Coefficient of variation in %.
[f] Prisms unacceptable for testing purposes.

Since the grain form of crushed stone aggregates deviates considerably from the spherical form and since the aggregates have a rough surface texture apart from this, it is more difficult to compact them, and they have a greater void space. In the case of the same sieve line and similar consistency, the quantity of binder which they require is higher by 10–20 vol %.

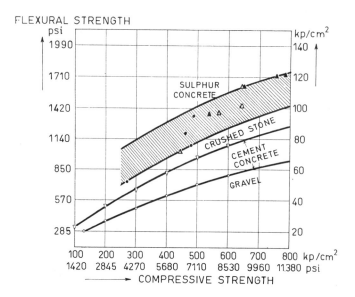

Figure 7. Comparison of the ratio between the flexural strength and the compressive strength of cement concrete and sulfur concrete

On the other hand, crushed stone concrete has a remarkable advantage. Because of good compaction and linking, the sharp grains form a closer matted structure and thus have a better inner binding. For the same reasons they also adhere better in the concrete than spherical and smooth pebbles. When the grain distribution was the same, the sulfur concrete had a compressive strength which was 32% higher than with crushed stone aggregate.

If one compares the ratio of flexural to compressive strength for sulfur concrete and cement concrete, further interesting aspects appear (Figure 7). If the ratio of flexural strength to compressive strength of cement concrete is 1:10 to 1:8 according to type of aggregate, the ratio for sulfur concrete is 1:6. For modified sulfur concrete still better ratios can be achieved as is shown below. Favorable, and thus low-strength, ratios are important in the case of very heavy loading.

Influence of Aggregate Basicity on the Strength of Sulfur Concrete. Up to now, it has not been determined what gives binder character to elementary sulfur. It seems improbable that only chemical bonds are responsible for this, since sulfur concrete can be melted again and again without a reduction in strength. In order to state influences in the micro-range, examinations with a scanning electron microscope and an x-ray photoelectron spectroscope were carried out.

Scanning Electron Microscope. A scanning electron microscope, equipped with a wave-length dispersive spectrometer, was used to examine the possibility of preferred accumulation of sulfur to certain mineral components. In the SEM technique a finely bundled electronic ray scans the sample by way of a screen. With the spectrometer, the emitted x-ray line spectrum can be used to determine the element distribution. The share of a chosen element on the surface appears in the form of light points in statistical distribution. Sulfur concrete samples were coated with a gold layer which was 500 Å thick. Figure 8 compiles x-ray distribution pictures of sulfur, silica, aluminum, calcium, potassium, and iron. The light zones in the individual pictures show the respective elements semiquantitatively. The photos are based on a crushed stone granulit bound with sulfur. This figure shows that sulfur does not accumulate preferentially to certain mineral parts.

X-ray Photoelectron Spectroscopy (ESCA). In order to state changes in the binding energies, ESCA (electron spectroscopy for chemical analysis) examinations were carried out. The respective kinetic energy of electrons which a substance emits when x-rays are encountered is measured. The bond energy can be determined by the Einstein relation and by considering the work function. According to the respective chemical environment of an atom under consideration, electrons can experience small changes on their inner orbitals which influence the binding energy.

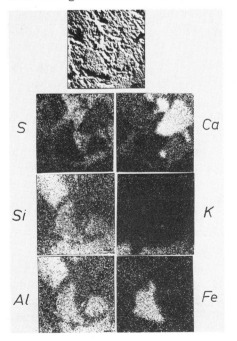

Figure 8. X-ray distribution photographs of a crushed stone granulit bound with sulfur

In the case of an oxidation, for instance, the remaining electrons were bound more firmly to the nucleus. This is reflected in a higher ionization energy and consequently in a lower energy of the photoelectrons.

In the case of iron, sodium, silica, and oxygen no changes in the binding energies could be stated. With aluminum, a shift of the binding energies up by about 1 eV compared with the pure mineral material was observed in the sulfur concrete samples. The binding energy of the sulfur in the concrete samples, on the other hand, was shifted down about 1 eV compared with pure sulfur. As the shifts which were found are much too small for a chemical bond, possible polarizations could have caused them. However, these are probably not responsible for the binding character of sulfur.

Modified Sulfur Concrete

Since plasticized sulfur has better mechanical properties than elementary sulfur, the properties of sulfur modified with styrene or dicyclopentadiene (DCP) as binder for aggregates were examined. These modifiers seemed to be the most favorable because of both improvement in properties and low price. Furthermore, as stated in the literature, DCP-modified sulfur has a lower combustibility (15).

Granulit and basalt were used as aggregates. The grain composition was based on sieve line 4 and 5. The respective quantities of sulfur corresponded to those which were used with unmodified sulfur concrete mixtures (16).

The addition of plasticizers to sulfur melts often results in big changes of the viscosity. Since with large increases in the viscosity of the sulfur melt the workability may be influenced unfavorably, we first determined quantitatively the changes in the viscosity of sulfur melts as a function of addition of the modifier and of reaction time.

Sulfur Concrete Modified with Styrene. If styrene is added to sulfur melts, the viscosity rises to a maximum within a few minutes irrespective of the styrene concentration (Figure 9). After reaching the maximum value, the viscosity decreases steadily to a constant final value which depends on the amount of styrene added. This characteristic behavior may be caused by thermal polymerization of the styrene followed by degradation of the polymer by reaction with sulfur.

The use of styrene to modify sulfur melts and thus to improve the binder properties was unsuccessful. When producing sulfur binder modified with styrene, reaction times of at least 1.5 hr at 140°C were used, since by then the viscosity maximum had been passed and a constant viscosity value had been reached. Sulfur melts plasticized by

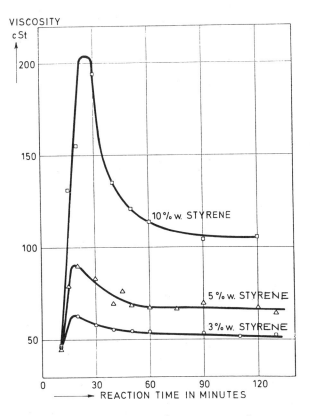

Figure 9. Viscosity of a sulfur–styrene melt vs. reaction time and styrene content. Reaction temperature, 140°C.

styrene have a comparatively high melt viscosity, making homogeneous mixing and processing especially difficult.

In examining the strength properties of sulfur concrete modified with styrene, the compressive strength was insufficient and worse than in the case of unmodified sulfur concrete (Table VI). After 70 days storage, however, the compressive strength of the test samples modified with 10 wt % styrene increased to 432 kp/cm², which is an average of 46%. The increase in the test samples modified with 5 wt % styrene was 4%, corresponding to a value of 526 kp/cm².

Sulfur Concrete Modified with Dicyclopentadiene (DCP). DCP was a better plasticizer for improving the strength of the binder. When DCP is used to modify sulfur, one can state the following quantitative viscosity changes depending on the reaction temperature, the reaction time, and the concentration of DCP (Figures 10a and b). Optimal

Table VI. Strength Test Results of

Sample No.		Density		Flexural Strength			
		Results (kg/dm^3)	Average (kg/dm^3)	Results (kp/cm^2)	$\overline{X}$ (kp/cm^2)	Sx[b]	Vx[c]
205/1	100 T Schwefel	2.28		127			
	5 T Styrol	2.24		110			
		2.26	2.26	140	125.7	15.0	11.9
205/2	100 T Schwefel 10 T Styrol	2.26		80			
		2.29		108			
		2.26	2.27	85	91	14.9	16.4

[a] Aggregate, granulit; test age, 24 days.
[b] Standard deviation.

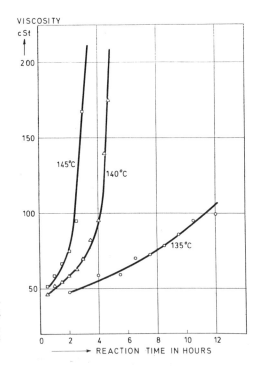

Figure 10a. Viscosity of a sulfur–DCP melt vs. the reaction temperature, the reaction time, and the DCP content. Reaction temperature, 140°C.

Styrol-Modified Sulfur Concrete[a]

Compressive Strength				
Failure Load		$\overline{X}$		
(Metric Tons)	*(kp/cm²)*	*(kp/cm²)*	Sx[b]	Vx[c]
12.0	480			
11.25	450			
13.1	524			
13.5	540			
12.7	508			
13.4	536	506	35.2	6.9
7.0	280			
7.3	292			
9.2	368			
6.6	264			
6.6	264			
7.7	308	296	39.1	13.2

[c] Coefficient of variation in %.

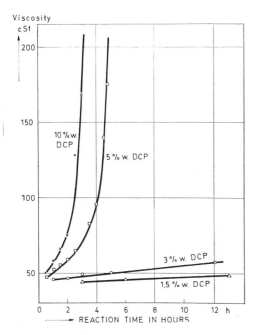

Figure 10b. Viscosity of a sulfur–DCP melt vs. the reaction temperature, the reaction time, and the DCP content. Reaction temperature, 140°C.

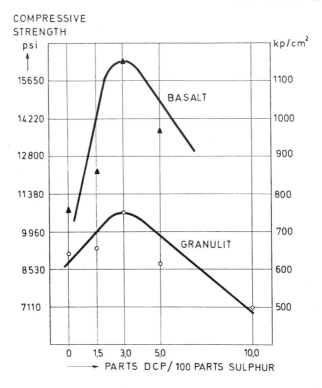

Figure 11. Compressive strength of sulfur concrete samples made with basalt and granulit aggregate vs. the DCP content

processing is guaranteed when the viscosity of the sulfur–DCP melt is 50–70 centistoke.

The influence of the DCP content in the sulfur binder on the compressive strength is represented in Figure 11. The comparison covers sulfur concretes with 1.5, 3,0, 5.0, and 10 wt % DCP in the binder. The highest values for compressive strength were obtained for both granulit and basalt sulfur concrete with 3 wt % DCP in the binder. The highest compressive strength obtained with DCP-modified sulfur concrete amounted to 1150 kp/cm². For comparison, normally constructed cement concrete has a compressive strength of about 160 kp/cm².

The changes in the flexural strength are extraordinary. As shown in Figure 12, the flexural strength rises proportionally with a rising DCP content. While unmodified sulfur concrete has a ratio of flexural strength-to-compressive strength of about 1:6, this ratio is about 1:2 for sulfur concrete modified with 10 wt % DCP. According to the expected loading, certain strength ratios may be adapted for DCP-modified sulfur concrete (Table VII).

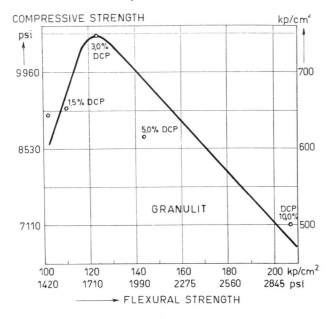

Figure 12. Compressive strength vs. flexural strength of sulfur concrete samples with different DCP content in the binder

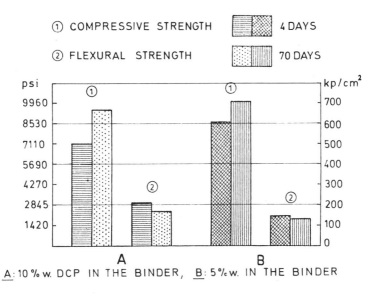

Figure 13. Influence of the storage time on the compressive and flexural strength of DCP-modified sulfur concrete samples. (A) 75.9 wt % graulit, 21.9 wt % sulfur, 2.2 wt % DCP. (B) 76.8 wt % graulit; 22.1 wt % sulfur; 1.1 wt % DCP.

Table VII. Strength Test Results

Sample No.		Density Results (kg/dm^3)	Density Average (kg/dm^3)	Flexural Strength Results (kp/cm^2)	$\overline{X}$ (kp/cm^2)	Sx^c	Vx^d
$205/3^a$	100 T Schwefel 10 T	2.36		210			
					207.5	3.5	—
	DCP	2.37		205			
		2.35	2.36	—			
$205/4^a$	100 T Schwefel 5 T	2.36		140			
	DCP	2.35		140			
		2.33	2.35	149	143	5.2	3.6
$205/5^a$	100 T Schwefel 3 T	2.39		133			
	DCP	2.36		122			
		2.38	2.38	114	123	9.5	7.7
$205/6^a$	100 T Schwefel 1.5 T	2.39		110			
	DCP	2.38		112			
		2.39	2.39	106	109	3.1	2.8
$307/7^b$	100 T Schwefel 5 T	2.44		218			
	DCP	2.46		210			
		2.43	2.44	216	215	4.2	1.9
$307/8^b$	100 T Schwefel 3 T	2.45		172			
	DCP	2.46		158			
		2.47	2.46	163	164	7.1	4.3
$307/9^b$	100 T Schwefel 1.5	2.49		125			
	DCP	2.48		136			
		2.48	2.48	121	127	7.8	6.1

[a] Aggregate, granulit; test age, 4 days.
[b] Aggregate, basalt; test age, 4 days.

of DCP-Modified Sulfur Concrete

Failure Load		Compressive Strength		
(Metric Tons)	*(kp/cm²)*	$\overline{X}$ *(kp/cm²)*	Sx[c]	Vx[d]
12	480			
12.5	500			
12.3	492			
13.0	520	498	16.8	3.4
—	—			
14.0	560			
15.1	604			
15.6	624			
15.9	636			
15.5	620			
15.6	624	612	27.2	4.5
19	760			
18.8	752			
17.8	712			
18.2	728			
19.0	760			
19.2	768	747	21.9	2.9
16.35	654			
16.45	658			
15.3	612			
16.0	640			
16.85	674			
17.0	680	653	24.7	3.78
24.5	980			
25.0	1000			
23.5	940			
24.0	960			
24.5	980			
24.2	968	971	20.5	2.1
26.2	1048			
27.4	1098			
29.5	1180			
30.0	1200			.
29.9	1196			
29.8	1192	1152	64.1	5.5
20.25	810			
21	840			
19.75	790			
21.75	870			
23	920	862	59.8	6.9
23.5	940			

[c] Standard deviation.
[d] Coefficient of variation in %.

As with sulfur concrete modified with styrene, there were also changes in strength with the sulfur concrete modified with DCP which depended on the plasticizer content (Figure 13). These changes in strength can be attributed to the unreacted sulfur, part of which in the beginning is in the amorphous form and which recrystallizes proportionally with increasing storage time and reinforces the polymer matrix. After 70 days storage at room temperature the compressive strength increased by 35%, and the flexural strength decreased by 18% for the sulfur concrete samples whose binder was modified with 10 wt % DCP. In the cases where the binder was modified with 5 wt % DCP, the average change in compressive and flexural strength was a 15.5% increase and a 5.6% decrease, respectively, in the same storage period.

Finally some important properties of cement concrete, polymer concrete, and sulfur concrete are compared with each other. For the polymer concrete the data of a commercial polyester concrete were used.

As shown in Table VIII, about the same compressive strength can be reached with DCP-modified sulfur concrete as with polyester concrete. The temperature loading lies in about the same range in both cases.

Table VIII. Comparison of the Important Properties of Cement, Polymer, and Sulfur Concretes

Concrete	Cement	Polyester	Sulfur	DCP–Sulfur
Composition (wt %)	68.2 crushed stone 22.8 cement 9.0 H_2O	90 gravel 10 polyester	79.7 crushed stone 20.3 sulfur	79.3 crushed stone 20.1 sulfur 0.6 DCP
Specific gravity (kg/dm^3)	2.40	2.28	2.52	2.46
Compressive strength (kp/cm^2)	600	1200	760	1150
Flexural strength (kp/cm^2)	75	230	120	160
90% ultimate strength reached after	28 days	24 hr	10 hr	24 hr
Pore volume of the hardened concrete (vol %)	8–25		1–4	1–4
Thermal conductivity ($kcal/m\ hr\ °C$)	1.75		1.2	1.1
Stability against thermal loading (°C)	400–500	80	95	80
Combustibility	no	yes	yes	yes
Water absorption (wt %)	2	0.13	0.12	0.09

Table IX. Material and Fuel Cost to Produce 1 m³ of Different Concretes

	Price/ Unit	Cement	Poly- ester	Sulfur		DCP–Sulfur	
Materials costs (US $/ton)							
aggregate	4	6.6	8.2	8.1	8.1	7.8	7.8
sulfur	32.5	—	—	16.6	—	16.1	—
sulfur	70	—	—	—	35.8	—	34.6
cement	41.1	22.6	—	—	—	—	—
polyester	750	—	171	—	—	—	—
DCP	200	—	—	—	—	3	3
Fuel costs 0.12 US/10⁴ kcal		—	—	1.6	1.6	1.5	1.5
Total (US $/m³ concrete)		29.2	179.2	26.3	45.5	28.4	46.9

Comparing the cost of the various concretes shows that at times of low sulfur prices, sulfur concrete can even compete with high-strength cement concrete. The costs of polymer concrete per cubic meter are quite enormous. Yet the substantially higher production costs of polymer concrete have not been considered yet. The prices used for this comparison are based on European conditions. To assess the sulfur concrete costs we used two extreme sulfur prices which had to be paid in the last years. The total costs of DCP-modified sulfur concrete rose only negligibly with the addition of this plasticizer (Table IX).

Conclusions

Sulfur concrete has several important advantages compared both with cement concrete and with polymer concrete. Its quick development of strength; high mechanical loading; enormous chemical resistance; low wettability, water absorption and thermal conductivity; good freeze–thaw resistance; good resistance to abrasion; as well as the possibility of uncomplicated forming with little deviation from measure admit a broad spectrum of applications. The extremely short times for turn-over for the forming could be a great economic benefit in producing prefabricated concrete products. DCP-modified sulfur concrete should be preferred to unmodified sulfur concrete because of its more favorable mechanical, chemical, and physical properties.

Acknowledgment

This work was sponsored by the Austrian Department of Health and Environment Protection, by the Österreichische Industrieverwaltungs-AG, and especially by ÖMV-AG. The authors wish to thank F. J. Pass and

A. Ecker for their interest and many useful discussions. Furthermore the authors wish to thank J. Wernisch for carrying out the SEM and ESCA examinations.

Literature Cited

1. Wright, A. H., U.S. Patent **25,074** (1859).
2. Duecker, W. W., *Chem. Metall. Eng.* (1934) **41** (11), 583.
3. Duecker, W. W., *Min. Metall.* (1938) 473.
4. Payne, C. R., Duecker, W. W., *Chem. Metall. Eng.* (1939) **46** (12), 766.
5. Payne, C. R., Duecker, W. W., *Chem. Metall. Eng.* (1940) **47** (1), 20.
6. Dale, J. M., Ludwig, A. C., *Ci. Eng. ASCE* (1967) 66.
7. Frusti, R. A. J., *Mil. Eng.* (1967) **387**, 27.
8. Crow, L. J., Bates, R. C., *U.S. Bur. Mines Rep. Invest.* (1970) 7349.
9. Loov, R. E., Vroom, A. H., Ward, M. A., *Polym. Congr. Inst. J.* (1974) 86.
10. Malhotra, V. M., Mines Branch Investigation Rep. **73-18**, Jan. 1973.
11. *Ibid.*, **74-25**, Jun. 1974.
12. ÖNORM B 3304 (edition January 1969).
13. Grün, R., *Beton* (1937) 120.
14. Hummel, *Beton* ABC (1960) **60**, 12.
15. Dale, J. M., Ger. Off. Patent **2,341,302**, 1974.
16. Gregor, R., Hackl, A., Ecker, A., Pass, R., ÖGEW-DGMK Gemeinschaftstagung Salzburg Oktober 1976 Neue Verwertungsmöglichkeiten für Elementarschwefel.

RECEIVED April 22, 1977.

Stability of Sulfur-Infiltrated Concrete in Various Environments

J. A. SOLES, G. G. CARETTE, and V. M. MALHOTRA

Canada Centre for Mineral and Energy Technology, Department of Energy, Mines and Resources, Ottawa, Canada

Sulfur-infiltrated concrete has been shown to be more durable in most environments than untreated normal portland cement concretes. Long-term strength maintenance indicates that the concrete is stable indefinitely in ambient, dry-to-humid conditions and can endure extended exposure to cyclic freezing and thawing. However, sulfur is leached from the infiltrated concrete by aqueous media, which causes deterioration and may affect greatly its long-term durability. This unstable condition is related apparently to the presence of polysulfide anions which are formed during the infiltration process and are highly soluble in the alkaline pore solutions of wet concrete. The character of the infiltration and leaching products is described, and the reactions involved are discussed.

Production of sulfur in North America from sour natural gas and as a by-product of mining is expected to yield a surplus of 25×10^6 metric tons in Canada alone by 1980. Much attention therefore has been directed recently to developing new applications for this currently abundant resource. One such application is as a construction material. In Canada, most experimental work is being done by the Canada Centre for Mineral and Energy Technology (CANMET), the National Research Council, and the Sulphur Development Institute of Canada (SUDIC), which finances research through universities and commercial testing laboratories (1). Similar studies are being pursued in laboratories in the U.S. and elsewhere (2).

Research at the Industrial Minerals Laboratory of CANMET has included the development of a technique (3) for producing high-strength,

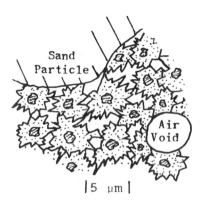

Figure 1. Schematic of hydrating cement grains, illustrating growth of C–S–H gel around them and construction of interconnected, water-filled interstices. Matrix is enlarged relative to scale.

sulfur-infiltrated concrete which could make commercial production of specialized pre-cast units economically feasible. The concrete, in addition to having high strength, appears to be highly durable under normal sub-aerial weathering conditions and able to resist deterioration in saturated freezing–thawing, acidic, and saline environments. However, sulfur-infiltrated concrete is not stable consistently, and until the causes of deterioration are known, its usefulness as a construction material will be limited at best. This chapter summarizes the results of durability studies and experiments undertaken at this laboratory to determine physical and chemical conditions which contribute to this deterioration.

Composition and Structure of Sulfur-Infiltrated Concrete

The Matrix of Uninfiltrated Concrete. Normal hydration of cement yields a rather complex assemblage of several phases, as described elsewhere (4). Because most of the hydrates are poorly crystallized, amorphous, or at least of colloidal dimensions, the physical structure of the matrix is not well known. In Neville (5) the model presented is a solid continuum of gel particles surrounded by a film of adsorbed water held in gel "pores," the mass being laced through with a feeder network of larger capillaries. Others (6) have proposed different structures for the gel. For example, the studies of Verbeck and Helmuth (7) suggest that aggregations of gel form in water-filled spaces surrounding cement grains, and a continuous maze of coarse and sub-micron size capillaries exists which provides water for their continued hydration. Perhaps the gel is simply an interlocking mesh of hydrated particles through which water migrates unimpeded to the cement interface or through which ions migrate outward.

Whatever the fine structure and reactions associated with hydrating cement paste, the permeability of the hardened matrix will depend on the sizes of interconnected capillary openings remaining after hydration.

Young's (8) model illustrates how the degree of hydration would affect their sizes. As hydration proceeds (Figure 1), the gel formed will constrict or bridge progressively the water-filled interstices until the process is terminated, either by exhaustion of hydratable cement or by removal of water to prevent further hydration. The permeability of a partly cured, dried concrete matrix will therefore be determined at the hydration stage by the amount of gel shrinkage caused by drying. The structure of the matrix would thus simulate that of a loosely knit mush of gel particles linked across the interconnecting pores or capillaries.

Structure of the Infiltrated Matrix. The basic structure changes little during infiltration unless reaction occurs; the liquid will simply fill the interconnected capillaries and voids, if conditions permit or compel it to do so.

Molten sulfur performs well as an infiltration liquid provided the proper conditions are maintained; some properties which affect its behavior are given in Table I. It wets hydrated cement phases when they have been dried, and its moderately low viscosity of about 10 cP at 130°C permits easy migration through and sealing of capillaries in the hardened paste. According to the model given in Figure 1, the matrix of fully infiltrated concrete would be a rigid, almost impermeable mass of hydrated cement particles bonded together with crystalline sulfur.

Preparation of Sulfur-Infiltrated Test Specimens

The process of infusing concrete with molten sulfur was recorded as early as 1924, when Kobbé (10) attempted to produce a strong, water-resistant material. The simple immersion method he used is effective if sufficient time is permitted to infiltrate precast units in controlled environments (11). However, mass production and field applications may demand that the procedure be accelerated, for example, by using modi-

Table I. Fundamental Properties of Sulfur

Allotrope	Structure	mp	sp gr	δV [a]
S_α	S_8 rings, orthorhombic	112°C	2.03	
S_β	S_8 rings, monoclinic	119°C	1.96	>3.8%
	Liquid		1.80	>8.8%

$$S_\alpha \xrightleftharpoons[\text{metastable}]{96°} S_\beta \xrightleftharpoons[115°]{119°} S_{liq.} \quad 159° \begin{matrix} \longrightarrow S_\lambda - (S_8 \text{ rings}) \\ + \\ \longrightarrow S_\omega - (S_X \text{ chains}) \end{matrix}$$

[a] Viscosity of liquid: 11 cP at 120°C; 7 cP at 159°C; 10^5 cP at 187°C. S_β inverts to S_α slowly at normal temperatures. Polymers (S_λ and S_ω) can be retained metastably by cooling rapidly (9).

fied evacuation and pressurizing techniques such as those used to impregnate concrete with viscous, polymerizable liquids (12).

Much of the CANMET investigation required that the strength characteristics of specimens from different mixes be similar, which would be obtainable only with complete infiltration of similar concrete. The procedures were standardized therefore to obtain the best matrix condition for infiltration within about 2 days. A water/cement ratio of ~ 0.7 was selected after much testing; it yields a highly permeable concrete which can be handled after 24 hr of moist-curing and from which uncombined water can be removed with 20 hr of drying. The infiltration process involves evacuation of the immersion vessels to expedite the displace-

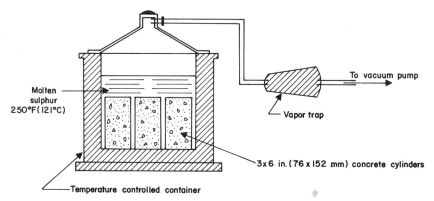

Figure 2.   Sulfur infiltration system used in this investigation

Table II.   Materials, Typical Mixes, and Procedures Used to Prepare Sulfur-Infiltrated Concrete Specimens

Materials

Cement: Portland, type 1; Sulfur: 99.9% pure, commercial
Aggregates: C.A.: Crushed igneous rock; F.A.: natural sand

Typical Mix Proportions

F.A./C.A. = 1:1   Agg./Cem. = 6.5:1   Cement = 200 Kg/m^3
Water/Cement ratio: 0.69 by weight

Procedures

1. Moist-cure concrete 24 hr; demold.
2. Dry specimens 20 hr at 130°C, weigh; test control specimens.
3. Immerse hot specimens in molten sulfur held at 125°C in vessels.
4. Evacuate system to ~ 7 mm Hg (1 kPa) for 2 hr.
5. Release vacuum; soak specimens for ½ hr at atmospheric pressure.
6. Remove specimens; drain off excess sulfur; cool in air.
7. Weigh to obtain sulfur loading; test control specimens.

ment of air from capillaries, followed by release of the vacuum to accelerate infiltration with atmospheric pressure. The system is illustrated in Figure 2, and the materials and procedures used are summarized in Table II.

Experimental Objectives and Tests

The investigation of sulfur-infiltrated concrete at this laboratory included three major areas of study: strength development and retention, resistance to freezing and thawing, and stability in various aqueous media. It was hoped that some conclusions could be drawn regarding practical manufacturing limitations, long-term durability in physically or chemically hostile environments, and potential usefulness of such concrete.

Strength Development in Sulfur-Infiltrated Concrete. Infiltration of a penetrable concrete matrix with any liquid which can solidify will increase the strength of the concrete in proportion to the amount of liquid introduced and to the strength of the bond formed between infiltrant and hydrated cement matrix. The infiltration techniques used at CANMET often yielded fully infiltrated concrete with compressive strengths exceeding 100 MPa, as much as 13 times the strength of uninfiltrated reference moist-cured specimens from the same mix, or three to four times that of concrete having a 28-day strength of 30 MPa (4500 psi). Splitting-tensile strengths increased by the same factor—high for a composite of weak concrete infiltrated with a brittle material. Much experimental work on the causes of strength increase had been done with polymers but not with liquids which crystallize. Therefore, some preliminary tests were made to determine how sulfur loading affected strength development and whether beneficial reactions occurred which might enhance it.

SULFUR LOADING. The strength of sulfur-infiltrated specimens produced in this laboratory varied greatly over a wide range of sulfur loading, supporting the evidence from previous studies that the volume of infiltrant is a principle factor affecting strength increase. The specimens in Figure 3 illustrate the fact clearly; the compressive strengths of the cylinders increased with penetration depth of the sulfur. The relation is less evident with more complete infiltration because a wide scatter in test results (Figure 4) was obtained. However, higher pressures may be necessary to fill the finer capillaries consistently, as Manning and Hope (*13*) showed in experiments with polymers.

MATRIX BONDING: INFRARED ABSORPTION STUDIES. It is possible that some form of reaction could occur between liquid sulfur and the hydrated cement phases to produce a chemical bond which would contribute to the strength of the composite, perhaps more noticeably with filling of the

Figure 3. Cylinders of infiltrated concrete showing decreasing depths of sulfur penetration, left to right. This may be caused by decreasing the water/cement ratio, by lengthening the period of hydration, by drying less, or by permitting less time for infiltration.

finer capillaries. This possibility was examined by obtaining infrared absorption analyses (14) of uninfiltrated and sulfur-infiltrated concrete mortar and by comparing the spectra. The IR absorption curves, reproduced in Figure 5, show distinct spectral intensity differences between

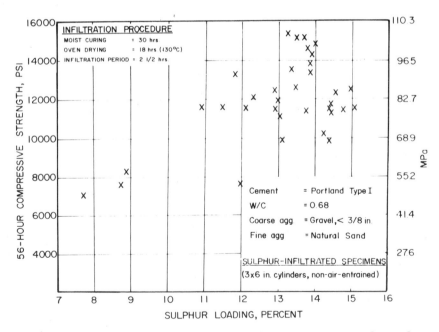

Figure 4. Relation between sulfur loading and compressive strength of sulfur-infiltrated concrete. Wide scattering at higher loadings suggests variable porosity and/or inconsistent filling.

uninfiltrated and infiltrated concrete; however, the types of bonding could not be established from these spectra. Enhancement of the 1080-cm⁻¹ frequency band suggests that SO_4^{2-} could have formed, but no corresponding bands had developed in the 580–670-cm⁻¹ range to confirm it. There is no evidence that H–S bonds had formed in the infiltrated specimens as might have been expected with hydrates present, particularly $Ca(OH)_2$; however, such bonds could have been destroyed

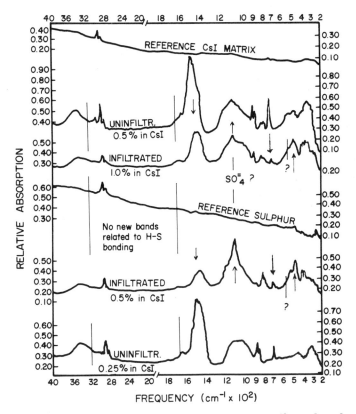

Figure 5. Infrared absorption spectra of uninfiltrated and sulfur-infiltrated concretes. Some distinct changes are arrowed.

during sample preparation. Finally, the distinct contractions of the 1435–1460-cm⁻¹ doublet and the 700-cm⁻¹ singlet cannot be explained.

The IR spectral data suggest that some changes in chemical bonding could have occurred during infiltration, possibly associated with modification of a hydrated cement phase, but no new bonds which might increase the concrete strength significantly were recognized in the spectra of infiltrated samples. It appears that most of the high strength obtained by infiltration has been produced by filling of capillaries and cross-linking of dried gel particles with sulfur.

Durability in Freezing–Thawing and Dry Environment. Long-term exposure testing of specimens to accelerated freezing in air and thawing in water has indicated that sulfur-infiltrated concrete resists destruction in such extreme environments remarkably well. Maintenance of specimen strength to beyond 1000 cycles was common, linear expansion seldom exceeded 0.08% after 1000 cycles, and ultrasonic pulse velocities changed little in most specimens. This high degree of durability reflects the low permeability of the concrete to water; indeed, some partially infiltrated specimens proved to be as resistant as others which were fully infiltrated.

Some specimens behaved poorly, developing wide cracks within 250 cycles. The weakness was related to stress cracks which had formed during demolding of the green concrete and which provided loci for rupture of sound concrete by ice wedging along the relatively weak, sulfur-filled planes (*see* Figure 7). The deterioration resembles that observed with sulfur concrete (15). Such occurrences pose two problems of particular interest to manufacturers—rough handling of green specimens must be avoided, to prevent damage, and highly permeable aggregates should be avoided, because infiltration with sulfur may not protect them effectively against freeze–thaw deterioration in wet environments. Strength tests of specimens stored in an open room showed that the original high strength of sulfur-infiltrated concrete is retained indefinitely in ambient dry-to-humid conditions.

Stability of the Infiltrated Concrete in Aggressive Media. The vulnerability of concrete to aggressive aqueous solutions depends principally on the permeability of the matrix to the solutions and on the reactivity of constituents with the ionic species being carried in solution. Treatments which decrease permeability, shield reactive phases, or otherwise reduce the reaction surface area will inhibit the potential activity of deleterious solutions.

Infiltration of the permeable concrete with molten sulfur yields a matrix which is almost impermeable to water. The freeze–thaw durability tests have indicated this is so, and immersion tests showed that fully infiltrated specimens absorbed less than 0.3% water by volume over several months, although methanol immersion and vacuum porosimeter measurements revealed that a pososity of over 5% was available for filling. A total shrinkage of about 13% (Table I) occurs when liquid sulfur crystallizes to the stable low-temperature S_α form, but much of the volume change appears to be accommodated in closed pores and intercrystal inversion fractures which affect the permeability little.

Because of the low permeability, chemical reactions with sulfur-infiltrated concrete should be confined principally to the surface of the concrete mass, and the deterioration rate will depend on how effectively sulfur films surrounding reactive hydrated cement phases in the matrix will shield them from active solutions. Several experiments were carried out to determine the stability of inflated concrete in different aqueous environments and to study the reactions involved.

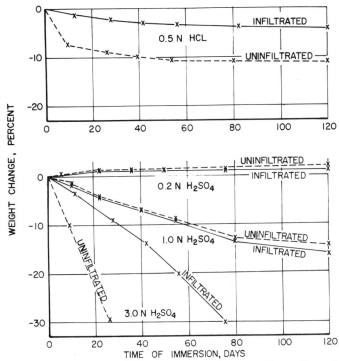

Figure 6. A comparison of leaching rates of sulfur-infiltrated and -uninfiltrated concrete cylinders by acids of different concentrations

Acidic Solutions. The digestion rates of uninfiltrated reference and infiltrated test specimens were measured in 0.5N HCl and in 0.2N, 1.0N, and 3.0N H$_2$SO$_4$ solutions. Both weight losses of specimens and changes in pH of solutions were measured at selected time intervals. The results of the tests, plotted in Figure 6, indicate that acid attack is inhibited by infiltration, but is clearly not prevented, particularly at higher acid concentrations, and that sulfuric acid is more destructive than hydrochloric because sulfates can form and disrupt reacting surface layers. In this regard it behaves like polymer-impregnated concrete.

Although these tests do support the long-established fact that infiltration of portland cement concrete with sulfur makes it more durable in acidic environments (10), they have revealed that reaction with the cement phases, especially Ca(OH)$_2$, proceeds nevertheless. Two of the specimens in Figure 7 illustrate the degree of penetration obtained over several months of digestion in the acids. No external changes were noticeable during immersion in HCl to show that internal reactions were taking place. Considering the model in Figure 1, one might speculate that the acidic solutions progress through the concrete via the hydrated phases, the rate of advance being controlled by the rate of diffusion of solutions and products through the unaffected sulfur maze. The conclusion which can be drawn from the experiment is that sulfur-infiltrated concrete may not be as durable in acidic solutions as has been assumed.

Figure 7.   Leaching of 3-in. cylinders of infiltrated concrete by aggressive media. (left) 1.5N HCl solution removed Ca(OH)$_2$ deeply, leaving a yellowish rim of sulfur and silicates; (right) 0.2N H$_2$SO$_4$ leaves a shallow, pitted rim containing CaSO$_4$ · 2H$_2$O; higher concentrations erode the surface more rapidly; (center) 1.3N NaOH has removed all sulfur from a deep rim, leaving only hydrates and aggregates.

The effect of pH on the stability of infiltrated concrete in acids is not yet known, pending a study with buffered solutions or partly dissociated acids. A leaching experiment using silage liquor with a pH of 4 has shown no effect after three months.

ALKALINE AND NEUTRAL SOLUTIONS. The instability of sulfur-infiltrated concrete in sulfatic soils has been long known (16), but the reason for deterioration is still poorly understood. Degradation of untreated portland cement concrete in such environments often appears related to the formation of gypsum or ettringite (17), but infiltration with sulfur should inhibit the lime–aluminate–sulfate reactions which produce these expansive compounds. It does, but poorly, which would imply that a reaction occurs which destroys the sulfur shield. To provide more information, several experiments were made to test the durability of sulfur-infiltrated concrete in aqueous alkaline and neutral conditions.

Tests in NaOH Solution. Specimens of infiltrated concrete were placed in 1.3N NaOH solution together with uninfiltrated specimens, examined and weighed at intervals, and finally broken in compression at 12 and 18 months. The only visible external change was a blotchy darkening, but specimen weights dropped steadily, and by the end of the period the strength had dropped to near the pre-infiltration level. Broken specimens showed that the sulfur had been completely leached from a broad band at the surface (Figure 7). When neutralized with acid, the slightly yellowish leach solution precipitated abundant sulfur and evolved H_2S. The experiment proved that sulfur is leached rapidly from infiltrated concrete when it is immersed in strongly alkaline media and therefore could not protect the hydrated cement phases from other destructive reactions.

Tests in Neutral Solutions. Prisms and cylinders of infiltrated concrete were placed in distilled water, and the system was permitted to stagnate. Within a few weeks, flowery exudations of sulfur began to appear on the surface, and sulfur was precipitated in the surrounding water, together with minor calcium sulfate and carbonate. Abundant calcium sulfate was also recovered from the water after evaporation. After removing the specimens from the liquid, strongly alkaline calcium- and sulfur-rich liquid globules began to extrude onto the dried surface in the vicinity of the sulfur exudations. The globules are apparently extrusions of the solutions which have leached sulfur from the interior of the specimens.

A similar experiment was run with expansion test prisms immersed in 2.5% Na_2SO_4 solution with the same results, although the reaction was slower. No significant prism length changes occurred during the immersion period.

These tests indicated that sulfur-infiltrated concrete still loses abundant sulfur when immersed in neutral and sulfatic solutions, even though the reaction is localized and relatively slow. Compression tests of a few cylinders showed there had been little reduction in strength over several months. Examination of sawn surfaces revealed no clearly leached zone, but the specimens had turned a mottled blue, except near the center. Faint peripheral fractures had developed, and coherence of the infiltrated matrix near the surface had decreased, suggesting that some leaching had taken place.

Tests in Concentrated Na_2SO_4 Solution. The study was extended to include a practical durability test of the concrete in the corrosive environment of a sodium sulfate plant, whose normal concrete foundations had deteriorated to the point where some needed replacing (Figure 8). Specimens were placed in areas where concentrated solutions of brine would wash over them intermittently, and where the temperature would

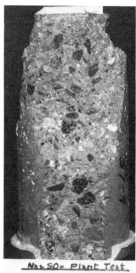

Figure 8. (left) Deteriorating portland cement concrete footings beneath evaporators in a sodium sulfate plant in Saskatchewan. (right) Fractured surface of sulfur-infiltrated concrete specimen placed in same area shows no deterioration after 9 months exposure.

Figure 9. Photograph of sulfur-infiltrated concrete cylinders immersed 10 months in distilled water. Note the abundance of flowery sulfur extrusions and the trace of a sulfur-filled fracture (↖) formed during demolding. Dark rings are iron sulfide.

occasionally drop below 0°C. Examination after 10 months revealed no sign of deterioration either exterior or interior, and compression tests of three specimens indicated that the strength of the concrete had increased slightly for no apparent reason. These exposure tests are being continued.

A Study of Leaching of the Infiltrated Concrete. The distilled water experiments provided the most information for examining the reactions involved in the leaching of sulfur-infiltrated concrete in aqueous media. Products selected from the specimens were analyzed, and the experiment was continued in a controlled environment. One specimen was partly immersed in distilled water under a nitrogen atmosphere. Fragments of the concrete were also sealed in a test tube.

ANALYSES OF THE EXTRUSION PRODUCTS. The photograph in Figure 9 shows the abundance of flowery sulfur exudations that may appear on the surfaces of specimens immersed in stagnant water. They grow mostly as a mound of prismatic crystals associated with minor gypsum but occasionally develop on pedestals as shown in Figure 10. X-ray

Figure 10. (left) Photograph of a flowery exudation of sulfur on concrete immersed in distilled water for 10 months. The basal mound of sulfur crystals is surmounted by a sulfur bloom grown at the tip of a calcite tubule (Magnification ×8). (right) Liquid extrusions on dried concrete surface adjacent to earlier sulfur exudation. Same scale.

diffraction and scanning electron microscope (SEM) analyses of one of the pedestals revealed that they are calcite tubules (Figure 11) which transported sulfur-rich solutions to the point of deposition.

The nature of the solutions carrying the sulfur from the interior of the concrete is revealed by the liquid which extrudes on the surface of

*Figure 11.   Scanning electron micrographs of a calcite tubule, (left) pro-
truding from surface of concrete in a mass of sulfur crystals (×44); (right)
showing crystal growth in the walls and striae in the conduit (×440)*

the concrete after a specimen is removed from water, and the surface is
dried (Figure 10). The liquid is a deep orange and highly alkaline; it
quickly develops a flexible skin which greatly retards evaporation,
gradually turns rubbery, and eventually solidifies. Only sulfur could be
detected by x-ray diffraction analysis of the dried yellow solid, but x-ray
energy dispersive analysis of a scanning electron microscope mount of
the liquid (Figure 12) showed that calcium is abundant also.

ANALYSES OF IMMERSION LIQUIDS AND PRECIPITATES. The experiment
under nitrogen provided vital information on the stability of the concrete–water system and the reactions taking place. Under a normal
atmosphere, sulfur leached from the infiltrated concrete was deposited
on the surface as described or was precipitated as a fine mud in the
immersion water, together with minor gypsum and calcite. Under nitrogen, however, the liquid turned yellow to pale orange, its pH rose to 11
or more, and precipitation ceased. When the solution was exposed to
air, it turned colorless and precipitated sulfur rapidly. This was accompanied by a pH drop to a normal 7.3 level.

THE FINAL ANALYSIS. The SEM analysis of the liquid extrusion
shows that sulfur is present in quantities much greater than the mono-
sulfide of calcium could provide. The most probable explanation is that
part of the sulfur is present in polysulfide anions, as the infiltration conditions appear to favor their formation (*18*). A sample of the yellow
solution obtained by leaching fragments under nitrogen was therefore
analyzed by laser Raman spectrometry to determine if the absorption

spectra of polysulfide anions were present. The analysis, reported elsewhere (*19*), revealed they were, although the species present were not identified.

Hypothesis of the Leaching Process. The results of these various analyses and observations permit formulation of a hypothesis of sulfur leaching from infiltrated portland cement concrete by aqueous media.

The leaching process requires the formation of polynuclear sulfide and accompanying thiosulfate anions by reaction of sulfur with a highly alkaline, aqueous medium. Certain cations appear to favor the reaction more than others; e.g., $Ca(OH)_2$ seems to activate reactions more than NaOH. Elevating the temperature to the boiling point accelerates the process but is not a necessary condition as the literature often implies. The hydrated lime in portland cement concrete provides the required environment when water is available to dissociate the hydroxide; the pH of the liquid is at or rises to the level where reaction can proceed, and sulfur is taken into solution. Probably the process begins during the infiltration stage, when reaction could occur between partly dried $Ca(OH)_2$ and the hot infiltrating sulfur, forming various polysulfide anions, some of which give the concrete matrix a distinct bluish cast (*20*). Their presence does not appear to be necessary for later leaching reactions, however.

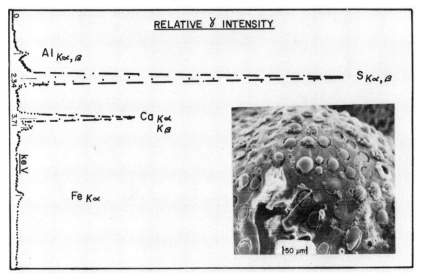

Figure 12.   Scanning electron micrograph of a liquid extrusion on infiltrated concrete removed from water and the corresponding x-ray energy dispersive spectrum showing that sulfur predominates over calcium. Globular masses erupted during evacuation.

Immersing dried infiltrated specimens in aqueous solutions regenerates the leaching conditions, and the process continues. Any polysulfides formed during infiltration will dissolve as liquid moves into accessible capillaries, then will migrate down the concentration gradient to the surrounding medium. The leach solution becomes more alkaline as $Ca(OH)_2$ is dissolved and reacts with adjacent sulfur. It is also possible that higher polysulfides are formed by the reaction of sulfur with lower sulfide species.

If the immersion medium is strongly alkaline, the anions will remain in solution, but if it is weakly alkaline or neutral, dissociation of a polysulfide will occur as it departs the concrete, and sulfur will precipitate either on the concrete surface or in the surrounding liquid. In either case the concrete should be leached of sulfur, but the process is much slower with neutral media, possibly because diffusion of leachate and polysulfide is slowed as sulfur precipitates near the surface.

Table III.   Possible Chemical Reactions Involving Sulfur-Infiltrated Portland Cement Concrete in Aqueous Media

Infiltration and Leaching

$$3\,Ca(OH)_2 + \begin{matrix} 12S \\ 10S \\ 8S \\ 4S \end{matrix} \rightleftarrows CaS_2O_3 + \begin{matrix} 2\,CaS_5 \\ 2\,CaS_4 \\ 2\,CaS_3 \\ 2\,CaS \end{matrix} + 3\,H_2O$$

$$Ca(OH)_2 + CaS_2O_3 + 4S \rightarrow CaS_5 + CaSO_4 + 2\,H_2O$$

$$CaS + 2\,S \rightarrow CaS_3$$

Precipitation

$$2\,CaS_5 + 3\,O_2 \rightarrow 2\,CaS_2O_3 + 6\underline{S}$$

$$2\,CaS_5 + H_2O + CO_2 \rightarrow Ca(SH)_2 + 8\underline{S} + \underline{CaCO_3}$$

(and other polysulfides ....)

$$CaS_2O_3 \rightarrow CaSO_3 + S$$

$$4\,CaS_5 + 3\,H_2O \rightarrow CaS_2O_3 + 3\,Ca(SH)_2 + 12\underline{S}$$

$$CaS_5 + 2\,H_2O \rightarrow Ca(SH)_2 + Ca(OH)_2 + 3\underline{S}$$

Dissociation of the polysulfides yields an alkaline solution which, together with dissolved $Ca(OH)_2$, increases the pH of the surrounding medium to the point where incoming polysulfide anions can remain in solution. If the system is closed or under an inert atmosphere, precipitation will cease, and the concentration of polysulfides in the surrounding liquid will gradually increase as sulfur is removed from the concrete.

Under a normal atmosphere, however, the anions are dissociated by oxygen and carbon dioxide, and precipitation of sulfur continues in the manner described, together with some calcite and gypsum.

Drying the surface of specimens briefly imposes a condition in which polysulfide solutions could move out of the concrete, possibly under osmotic pressure, in a state similar to that within the capillaries. Reactions would cease quickly.

The various reactions envisaged are summarized in Table III. The compounds shown are intended only to represent phases identified and ions which are suspected or could be in solution; the equations should not be construed as accurate or complete.

Conclusions

The investigation at CANMET has revealed much about sulfur-infiltrated portland cement concrete that must be considered prior to its use as a construction material. Its greatest potential is no doubt in the pre-cast industry, because the matrix can be preconditioned to the optimum state for infiltration. High-strength units can be produced rapidly on a commercial basis if provision is made for handling semi-hydrated weak concrete; also, they can be produced more slowly from well hydrated or high-strength concrete if a higher pressure is applied or a longer infiltration period is used during infiltration.

The inherent instability of sulfur-infiltrated concrete in aqueous media illustrated in this study may be the most important factor in utilization, because it will affect long-term durability of the concrete in many natural settings. The $Ca(OH)_2$ produced by the hydration of portland cement is a principal reactant in the leaching process, and while it remains sulfur could be extracted, leaving the matrix vulnerable to other destructive processes. The removal rate of sulfur will vary greatly, depending mostly upon the pH of the immersion medium; thus, the concrete deteriorates in alkaline sulfatic soils but is relatively stable in the corrosive neutral sulfatic solutions from the sodium sulfate plant.

Strongly acidic solutions below pH 2.5 attack $Ca(OH)_2$ and possibly other phases of infiltrated concrete, but weakly acidic solutions above pH 3.5 appear to have little effect.

No significant leaching of sulfur occurred in the freezing–thawing experiments, except locally where specimens were damaged during demolding and are more permeable. In the freeze–thaw environment, and possibly that of the sulfate plant as well, mobility of the immersion liquids may have prevented the pH from rising sufficiently at the concrete–liquid interface to permit formation or removal of polysulfides.

Sulfur-infiltrated concrete therefore may remain stable indefinitely in aqueous media at pH 3–7, and it may resist deterioration for long periods at pH 7–8.5 if the media are mobile. At either end of this range one should expect deterioration.

Acknowledgments

Contributions to the study were made by many scientific and technical members of this laboratory. In particular, we are grateful for the assistance of W. A. MacDonald for sample preparation and of D. M. Farrell and E. E. Berry for the research on reactions. The Sulphur Development Institute of Canada kindly provided monetary support for travel and the preparation of test specimens.

Literature Cited

1. "A Canadian Response to the Sulphur Challenge," Sulphur Development Inst. Canada, Summary Report of Progress, 1975.
2. Fike, H. L., "Some Potential Applications of Sulfur," ADV. CHEM. SER. (1972) 110, 208–224.
3. Malhotra, V. M., Soles, J. A., Carette, G. G., "Research and Development of Sulphur-Infiltrated Concrete at CANMET, Canada," Proc. Int. Symp. New Uses Sulphur, Pyrite, Madrid (1976) 184–201.
4. Brunauer, S., Greenberg, S. A., "The Hydration of Tricalcium Silicate and Dicalcium Silicate at Room Temperature," Proc. Int. Symp. Chem. Cem., 3rd (1960) 1, 135–165.
5. Neville, A. M., "Properties of Concrete," p. 22–35, Pitman and Sons, London, 1963.
6. Greene, K. T., "Early Hydration Reactions of Portland Cement," Proc. Int. Symp. Chem. Cem., 4th (1960) 1, 371.
7. Verbeck, G. J., Helmuth, R. H., "Structures and Physical Properties of Cement Paste, Proc. Int. Symp. Chem. Cem., 5th (1968) 3, 2–4.
8. Young, J. F., "Capillary Porosity in Hydrated Tricalcium Silicate Pastes," Powder Technol. (1974) 9, 173–179.
9. Currell, B. R., Williams, A. J., "Thermal Analysis of Elemental Sulphur," Thermochim. Acta (1974) 9, 255–259.
10. Kobbé, W. H., "New Uses for Sulfur in Industry," Ind. Eng. Chem. (1924) 16 (10), 1016–1028.
11. Smith, R. H., Shah, S. P., Naaman, A. E., "Investigations on Concrete Impregnated with Sulfur at Atmospheric Pressure," Univ. of Illinois College of Engineering; Report 76-1, 1976.
12. "Polymers in Concrete," Proc. Int. Conf. Polym. Concr., 1st (1975).
13. Manning, D. G., Hope, B. B., "The Effect of Porosity on the Compressive Strength and Elastic Modulus of Polymer Impregnated Concrete," Cem. Concr. Res. (1971) 1, 631–644.
14. Farrell, D. M., "Examination of Sulphur-Impregnated Concrete by Infrared Spectroscopy," Canada Centre for Mineral and Energy Technology; Investigation Report MSL 76-28(IR), 1975.
15. Malhotra, V. M., "Mechanical Properties and Freeze-Thaw Resistance of Sulphur Concrete," Canada Dept. Energy, Mines and Resources; Mines Branch Report IR 74-25, 1973.

16. Bates, P. H., "The Use of Sulphur in Rendering Cement Drain Tile Resistant to the Attack of Alkali," *Ind. Eng. Chem.* (1926) **18** (3), 209.
17. Mehta, P. K., "Mechanism of Expansion Associated with Ettringite Formation," *Cem. Concr. Res.* (1973) **3**, 1–6.
18. Auld, S. J. M., "The Reaction Between Calcium Hydroxide and Sulphur in Aqueous Solution," *J. Chem. Soc.* (1915) **107**, 480–495.
19. Berry, E. E., Soles, J. A., Malhotra, V. M., "Leaching of Sulphur and Calcium from Sulphur-Infiltrated Concrete by Alkaline and Neutral Aqueous Media," *Cem. Concr. Res.* (1976) **7**.
20. Chivers, T., Drummond, I., "The Chemistry of Homonuclear Sulphur Species, *Chem. Soc. Rev.* (1973) **2** (2), 233–248.

RECEIVED April 22, 1977.

6

Characteristics of Some Sulfur-Bonded Civil Engineering Materials

J. E. GILLOTT, I. J. JORDAAN, R. E. LOOV,
N. G. SHRIVE, and M. A. WARD

Department of Civil Engineering, The University of Calgary,
Alberta, Canada T2N 1N4

Sulfur can be used in mortars, concretes, and asphaltic paving materials. Based on test data on various sulfur mortars and concretes, the following physical properties are discussed: brittleness, strength, modulus of elasticity, thermal expansion, thermal conductivity, creep, and resistance to water and temperature cycles. Some problems, such as durability and the effect of H_2S on strength, have been overcome to a great extent. Sulfur-modified asphaltic concretes have potential advantages over their unmodified asphalt counterparts because of their higher resilient moduli and better durability and aging. For some applications sulfur-bonded materials are superior to presently used construction materials.

In Canada the stockpile of sulfur has grown to about 19 million tons in 1977, and supply presently exceeds demand by about 2–3 million tons per year (1). Sulfur is produced principally in Alberta as a by-product of the desulfurization of sour natural gas. The capacity of the plants is rated at 9 million long tons per year (2), and sulfur is being produced at an annual rate of about 6–7 million tons (Table I). It is predicted that this rate will be maintained for the next 5–8 years. Future production in Canada depends principally on the rate of development of the oil sands deposits in Alberta (3) and on whether new sources of sour gas are discovered; the latter could result from success in the exploratory deep-drilling program in the Rocky Mountain foothills of Western Canada. Supplies of gas discovered so far in the Arctic are not sour although there are deposits of elemental sulfur in the Sverdrup basin (4).

0-8412-0391-1/78/33-165-098$05.00/0

In 1975 world production of elemental sulfur exceeded 32.5 million tons (1). Exploitation by the Frasch process of the large deposits found in association with the salt domes of the Gulf Coast regions has made the United States the world's leading producer (5). In the U.S. a potential (although unlikely) production of 40 million tons of sulfur per year from all sources by the year 2000 has been forecast, compared with a forecast consumer demand of only 30 million tons (6). Greater use of oil sand, oil shale, and coal, again with sulfur extraction, holds further promise of increases in the supply of sulfur in western Canada, the U.S., and elsewhere.

Apart from the U.S., Mexico, and Canada, other major producers of sulfur are the U.S.S.R., Poland, and France (Table I). The countries which border the Persian Gulf have enormous reserves of sulfur in their deposits of sour petroleum and sour natural gas. Up to now these re-

**Table I.   Cement, Steel, and Sulfur Production
in Various Countries (3)**

Mineral by Country	1971	1972	1973
Hydraulic cement*ᵃ			
Canada	9,066	9,976	10,884
U.S.	80,316	84,556	87,498
France	31,910	33,339	33,863
Poland	14,420	15,417	17,143
U.K.	19,508	19,894	22,037
U.S.S.R.	110,596	114,970	120,703
Japan	65,515	73,120	86,007
Raw steel*ᵃ			
Canada	12,170	13,073	14,755
U.S.	120,443	133,241	150,799
France	25,197	26,515	27,849
Poland	14,041	14,855	15,496
U.K.	26,647	27,912	29,405
U.S.S.R.	132,979	138,438	144,403
Japan	97,617	106,814	131,535
Sulfurᵇ			
Canada	4,720	6,839	7,290
Mexico	1,161	929	1,583
U.S.	8,620	9,240	10,021
France	1,773	1,703	1,775
Poland	2,670	2,881	3,485
U.K.	43	40	28
U.S.S.R.	3,642	3,838	4,085
Japan	403	491	670

ᵃ Thousand short tons.
ᵇ Thousand long tons.

Minerals Yearbook

sources have had limited development, but if major petrochemical plants, refineries, and gas liquefaction plants are constructed as planned, it seems inevitable that the Near East will emerge as a major world supplier of sulfur.

Traditionally sulfur is used as sulfuric acid in the manufacture of fertilizers as well as plastics, paper products, paints, nonferrous metals, explosives, and many other minor products. The present large stockpiles together with the strong probability of even greater oversupplies in the future have stimulated increased research into new uses for sulfur.

Because of the large supply of sulfur, there is increased interest in its possible use in the construction industry (7–13). This chapter reviews research at The University of Calgary concerned with sulfur in civil engineering applications. Large volumes of materials are required for construction. The amount of sulfur which is available may be compared with the consumption of some of the principal construction materials (Table I). In Canada the annual production of sulfur is already a sizeable fraction of the yearly consumption of some of these materials. For example the annual sulfur production is about half that of raw steel and about three quarters that of portland cement. Elsewhere sulfur production is much smaller than that of presently used construction materials, but there are indications that sulfur production will be increasingly important.

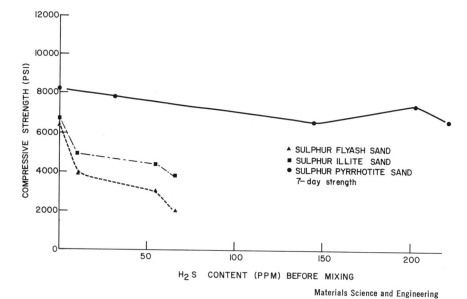

Materials Science and Engineering

Figure 1. Variation in compressive strength of sulfur mortars with hydrogen sulfide content before mixing (18)

BEFORE IMMERSION IN WATER

3 HOUR IMMERSION IN WATER

Figure 2. Effect of water on specimens containing 25% expansive clay by volume. Disintegration time depends on volume of clay present; as little as 1/2% by volume of clay will cause failure.

Construction Materials

Sulfur has been used for over a century as a bonding agent or additive in construction materials (*14, 15, 16, 17*). Within a few hours sulfur mortars and concretes have strengths which portland cement-bonded materials attain only after several weeks of moist curing. Sulfur has a low viscosity in the temperature range from 119° to 159°C, but above 159°C it becomes very viscous because of polymerization. Temperatures below the polymerization temperature were used in the present work to keep costs down and to facilitate mixing. Concretes were made by preheating the coarse aggregate to 125°–130°C to melt the sulfur which was added as a solid. The concretes contained 20.2% sulfur, 29.3% sand, 43.9% coarse aggregate, and 6.6% fines by weight. Mortars were made by mixing sand and fines with molten sulfur in proportions of 50% sulfur, 30% sand, and 20% fines by volume. Various materials were used as fines including flyash, pyrrhotite, and powdered shale composed mainly of illite. Inclusion of a suitable proportion of fines helps prevent segregation and improves workability. The fines commonly have important additional effects on properties of concretes and mortars depending on their composition. For example, pyrrhotite has a beneficial effect on the strength of sulfur in the presence of H_2S (*18*) (Figure 1) whereas small amounts of montmorillonite in fillers or in aggregate lead to very poor durability of composites, mortars, or concrete exposed to moisture (*19*) (Figure 2).

Sulfur crystallizes as a monoclinic polymorph which on cooling to room temperature inverts to an orthorhombic form. The most characteristic molecule is an eight-membered, crown-shaped ring (S_λ or S_8^R — cyclooctasulfur), but solids composed of hexatomic sulfur and numerous forms of catenapolysulfur are known (20, 21, 22). The change from liquid to orthorhombic crystalline solid is accompanied by a density increase of 13%. Additional material has to be added to concrete specimens to fill voids which result from the corresponding decrease in volume. Scanning electron microscope studies of cast polycrystalline sulfur using the technique of relocating identical areas (23) have shown changes in grain boundaries and crystal morphology and development of microcracks within a few hours of hardening from a melt (Figure 3).

Both the compressive and tensile strength of sulfur test specimens are affected by a number of variables including:

(1) Age of specimen (8)

(2) Size of specimen (12, 19) (see also Figure 4)

(3) H₂S content (18) (see also Figure 1)

(4) Mix casting temperature and temperature history of specimen (25, 26, 27)

(5) Temperature of testing (19)

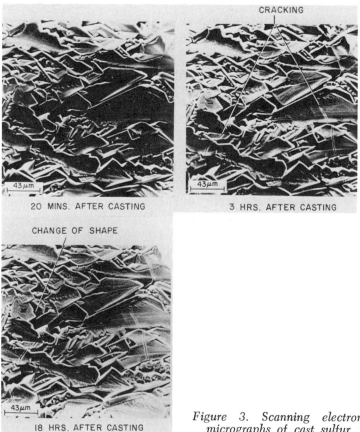

Figure 3. Scanning electron micrographs of cast sulfur

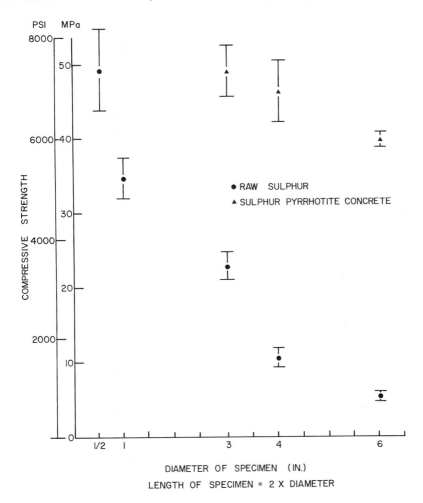

ASTM Journal of Testing and Evaluation

Figure 4. Variation in compressive strength with specimen size for raw sulfur and a sulfur pyrrhotite concrete; mean and standard deviation of six specimens (19)

(6) Mold material (*19*)—e.g., the seven-day strength of sulfur specimens cast in fiber molds is only 55% of the strength of specimens cast in steel molds (*19*).

The magnitude of some of these effects is reduced in sulfur-bonded concretes and mortars. Sulfur concrete mixes of the kind described fail in compression in a much more brittle manner than portland cement concretes of similar strength. This brittleness could effect performance; thus design methods used for structures made with portland cement concrete may be inapplicable.

The modulus of elasticity of raw sulfur was found to be 5.8 GPa (0.84×10^6 psi). Use of flyash as a filler caused a small decrease to 4.6 GPa (0.67×10^6 psi), but higher values were found for all other mixes. The mortars were somewhat less stiff than the concretes, the stiffest of which had a modulus of elasticity of 36.6 GPa (5.31×10^6 psi). These sulfur concretes have a stiffness similar to portland cement concretes of similar strength.

Raw sulfur has a high coefficient of thermal expansion. A value of $55 \times 10^{-6}/K$ at room temperature was found during the present work. If similar values were found in concretes, temperature differences between different parts of a single structural member would lead to high stress concentrations. Fillers reduce the value considerably; in sulfur mortars values varied between 21 and $58 \times 10^{-6}/K$ and between 11 and $29 \times 10^{-6}/K$ in sulfur concretes depending on the composition of the fillers and mix design.

Sulfur concretes creep more at room temperature than portland cement concretes, and deformation from this cause may increase rapidly with temperature (28). Excessive deflections of structural elements are an undesirable feature which could result from creep, but relief of residual stress is a potential benefit.

Durability failure of portland cement concrete results from such causes as freeze–thaw action, use of deicing agents, sulfate attack, seawater attack, alkali–aggregate reaction, wetting and drying cycling, and attack by acids (29, 30). Some of these causes of deterioration, such as alkali–aggregate reaction, do not affect sulfur concrete, and others such as attack by sulfates, deicing agents, many acids, and sugar have been greatly reduced in severity (31). However some agencies such as water (with aggregates and fillers containing swelling clays), freeze–thaw and thermal cycling, and bacteria pose greater threats to integrity, and there are unique problems such as the effects of heat and fire and corrosion of steel reinforcement. Results (19) indicate that by paying proper attention to the mineralogical composition of the aggregates and fillers and by incorporating suitable admixtures (32, 33), the effects of water and bacteria may be reduced or avoided.

The differences in properties between portland cement concrete and sulfur concrete suggest that the two materials will complement one another in civil engineering practice. The greater resistance of sulfur concrete to attack by sulfates and deicing salts makes its use probable in the marine environment, in bridge decks, and in curbs and gutters. The rapid strength gain and faithful reproduction of delicate surface features may well make the use of sulfur concrete attractive in certain precast operations. Sulfur concrete may be suitable for sewer pipes because of the resistance of the material to many forms of chemical attack provided additives are used to prevent degradation by bacteria (32, 33). Resistance to biological degradation is important in external applications since strong concentrations of sulfuric acid can be produced (32). Use in sewer pipes is commendable also since the risks of fire and damage from temperature cycling are low. Sewer pipes are buried deep enough so that temperature fluctuation from seasonal and diurnal changes is significantly damped (34).

Insulating Materials

In temperate and cold regions structures of nearly all kinds require good thermal insulation to reduce heating costs. In the construction industry, thermal insulators made from materials such as fiberglass and foamed plastics have been used widely because they are noncrystalline and incorporate a large void space with high entrapped air content.

Values of the thermal conductivity, k, have been determined in the present work with a thermal conductivity probe (24). It has long been known that sulfur has a low thermal conductivity although the values are even lower in such materials as PVC and expanded polystyrene. Sulfur-bonded composites made with inexpensive fillers such as soil and sand have thermal conductivities which are below those of typical portland cement concrete but with values higher than those of sulfur itself. The values for the composites are, however, still low as may be seen by comparison with the values for conductors such as steel and copper (Table II).

These results suggest that sulfur-bonded composites may have uses in civil enginering where thermal insulation is required. General principles indicate that increased voids content or a stable retention of a noncrystalline form by the sulfur would reduce the thermal conductivities of the composites. A composite incorporating potters flint did have a fine void structure, but nonetheless its thermal conductivity was higher than that of sulfur (24); presumably a still larger void content is required. Use of moist fillers such as damp soil produced foaming, but a stable, well distributed system of voids was not produced; the bubbles which formed collapsed under the conditions used.

Frost heave can be a major problem in maintaining roads and airfields and occurs with a particular combination of temperature, water supply, and type of soil. Frost-susceptible soils are generally silts and lean clays (35, 36). Also, in spring when near-surface ice melts, and the deeper soil remains frozen, an oversaturated subgrade can offer little or no structural support to the road. Foamed plastics limit frost penetration but offer little structural support. Preliminary results (37) from an instrumented road test section using a lightweight sulfur concrete suggest that highways may be designed to have good thermal insulating characteristics and sufficient strength to maintain integrity by bridging a thaw-weakened subgrade. Some sulfur-bonded clay soils develop high strengths and warrant further consideration for road construction (24) (see also Table II).

Table II. Strength and Thermal Conductivity of Some Engineering Materials

Material	Thermal Conductivity, k (W/m·K)	Compressive Strength
Expanded polystyrene, PVC, etc.	0.03–0.04	—
Sulfur	0.27	4000 psi 28 MPa (3″ Cyl.)
Sulfur-soils and lightweight aggregate sulfur concrete	0.3–0.6	2800–4800 psi 20–33 MPa
Asbestos–cement board	0.74	—
Concrete 1:2:4	1.37	5000 psi 35 MPa
Steel (1% C)	43	70,000 psi 480 MPa
Copper	386	33,000 psi 230 MPa

Surface Paving Materials

It has been known for many years that sulfur can modify the properties of asphalts (38). With the recent dramatic increases in oil prices and hence the price of asphalt, the use of sulfur in flexible paving materials is being examined with renewed interest (39–45), particularly in areas where sulfur is available and is less expensive than asphalt. In addition to the possible economic benefits, if asphalt can be partially or completely replaced by sulfur in flexible pavement concretes, other benefits in terms of improved material properties might also accrue.

There are, of course, many possible combinations of sulfur, asphalt, and aggregate. Knowledge of the mechanical properties, especially the time-dependent properties, of the many mixes produced to date is essential before engineers can foresee possible uses of these mixes. Many data on sulfur-modified asphalt paving materials have been generated. The work described here is part of a study (44) on sulfur-modified asphaltic concretes using ⅜-in. top-size heavyweight and lightweight paving aggregates.

The main study was aimed at developing a sulfur-modified asphaltic concrete which could be used with present manufacturing equipment. A Bencowitz-type sulfur asphalt emulsion (38) with a stabilizing additive was found to be promising. In our study (46) the effects of certain durability tests on some of the properties of two asphaltic concretes made with such an emulsion were compared with the effects of the same tests on concretes made with the unmodified asphalt. The durability tests were to determine the change in some low-temperature properties of the materials. In Calgary, the average air temperature in a recent period of six years (1969–1974 inclusive) was 3°C (38°F) within a range of −40°C (−40°F) to 35°C (95°F). Although the adsorption of radiated heat can increase the temperature of paved roads above the air temperature, the durability of paving materials at low temperatures (below 20°C) is clearly an important design factor in many North American communities and in the Calgary area in particular.

The resilient moduli of mixes containing ⅜-in. paving aggregate are shown as a function of temperature in Figure 5. Both concretes decrease in stiffness with an increase in temperature. The sulfur–asphalt bound mix has a slightly higher stiffness than the normal asphalt concrete at low temperatures. As the temperature increases, the decrease in stiffness is not as great for the sulfur–asphalt material as for the ordinary asphalt concrete. Thus the sulfur–asphalt concrete has a higher relative stiffness at the higher temperatures. Since for a given service strain criterion, the stiffer the material the thinner the necessary pavement layer thickness, it is clear that at low temperatures no significant saving is effected in the volume of material to be laid. However, at higher operating temperatures such a saving might be possible.

Vacuum saturating and freeze–thaw cycling has a much greater detrimental effect on the ordinary asphalt concrete than on the sulfur–asphalt concrete. Indeed after soaking, the sulfur–asphalt concrete shows a slight increase in stiffness, and the freeze–thaw cycling causes only a slight decrease in stiffness. The sulfur–asphalt concrete thus appears more durable than its ordinary asphalt counterpart.

The retention of stiffness after freeze–thaw cycling by the sulfur–asphalt-bound concrete with ⅜-in. aggregate is also shown in the second series of tests (Table III). Very little water adsorption was observed. The sulfur–asphalt bound material therefore shows better durability under the imposed conditions. However, neither concrete with the lightweight aggregate retained stiffness to the same degree. The sulfur–asphalt-bound concrete was again the stiffer, but lost a greater proportion of its stiffness during the test.

In the aging tests at −18°C (0°F), however, the sulfur–asphalt-bound concrete with lightweight aggregate certainly retains stiffness better than

Table III.   Results

Dry Values

Sample	Sample Type	$M_R{}^a$ (psi)	Density (lb/ft³)
1	⅜-in. agg/S–A binder	260×10^3	148.6
2	⅜-in. agg/A binder	111×10^3	146.6
3	L.W. agg/S–A binder	355×10^3	82.3
4	L.W. agg/A binder	73×10^3	78.9

[a] M_R = resilient modulus.

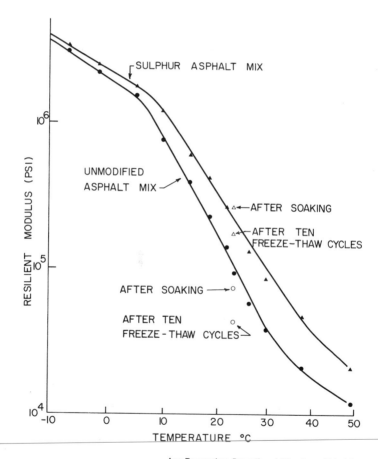

Low Temperature Properties of Bituminous Materials

Figure 5. Variation of resilient modulus with temperature, soaking, and freeze–thaw cycling for a sulfur-modified asphaltic concrete and a concrete in which the asphalt is unmodified (46)

of Test Series 2 (46)

Wet Values			*After 10 F-T Cycles*	
M_R (psi)	M_R as % of Dry M_R	% Water	M_R (psi)	M_R as % of Dry M_R
238×10^3	91.6	0.34	205×10^3	79.6
69×10^3	62.2	1.56	36×10^3	32.9
269×10^3	75.9	1.06	115×10^3	32.5
50×10^3	68.7	1.44	32×10^3	44.5

Low Temperature Properties of Bituminous Materials

Table IV.   Results of Storage at −18°C (46)

Sample Type	Wet/Dry	Original Density (lb/cu ft)	Final Density (lb/cu ft)	Original M_R ($\times 10^5$ psi)	Final M_R ($\times 10^5$ psi)	Final M_R as % of Original
1	Dry	148.2	148.1	3.86	2.79	72.3
		148.2	147.8	3.86	2.78	72.0
2	Dry	147.1	144.7	1.29	0.79	61.6
		147.6	145.0	1.29	0.71	55.3
1	Wet	148.0	147.7	4.02	2.67	66.4
		149.1	148.9	3.19	2.29	71.8
2	Wet	147.2	144.9	1.09	0.76	70.1
		146.7	145.8	1.55	1.10	71.4
3	Dry	82.1	78.9	3.90	3.49	89.4
		78.4	**75.4**	3.69	2.59	70.2
4	Dry	77.7	74.2	1.17	0.58	50.1
		80.9	76.0	1.13	0.66	58.2
3	Wet	76.0	73.2	4.96	2.11	42.5
		83.0	80.0	4.37	2.66	60.8
4	Wet	80.4	76.0	1.14	0.59	51.9
		76.5	a	1.05	a	a

a Specimen disintegrated in final resilient modulus test.

Low Temperature Properties of Bituminous Materials

the ordinary asphalt concrete (Table IV). The same effect is seen in the concretes with ⅜-in. aggregate. At the higher temperature (Table V) the sulfur–asphalt-bound materials again show less change in general than the ordinary asphalt concretes; the only exception being the ⅜-in. mix in the wet environment.

It is clear that sulfur–asphalt concretes have better potential durability than their non-sulfur counterparts. However, knowledge of the

Table V. Results of Storage at 60°C (46)

Sample Type	Wet/Dry	Original Density (lb/cu ft)	Final Density (lb/cu ft)	Original M_R ($\times 10^5$ psi)	Final M_R ($\times 10^5$ psi)	Final M_R as % of Original
1	Dry	148.8	146.7	3.72	4.19	112.7
		148.7	148.5	4.42	5.39	122.1
2	Dry	147.3	147.4	1.26	4.66	370.7
		147.5	147.3	1.17	3.69	315.3
1	Wet	148.9	145.9	3.69	0.70	19.1
		149.2	145.5	3.05	0.60	19.7
2	Wet	146.9 ^{a}	144.1	1.27	0.55	42.9
		147.1 ^{a}	144.2	1.32	0.54	40.8
3	Dry	79.6	77.3	3.94	5.67	143.9
		75.9	73.9	3.86	5.28	136.6
4	Dry	80.6	77.3	0.87	3.05	348.9
		79.6	76.9	0.93	3.06	328.5
3	Wet	76.2 ^{a}	75.5	4.10	1.43	34.8
		76.8 ^{a}	77.6	4.02	1.45	36.1
4	Wet	82.0	75.7	0.91	0.21	23.2
		78.8	74.7	1.13	0.52	45.9

Low Temperature Properties of Bituminous Materials

^{a} Final "dry" weight was higher than original "dry" weight.

fatigue properties of sulfur–asphalt concretes is necessary before the potential economic benefits may be defined more precisely.

Conclusions

There are strong indications that world production of sulfur will soon greatly exceed the demand from traditional users. In Canada the production rate of recovered sulfur is already approaching that of materials which are consumed in bulk such as portland cement. Hence suggestions that sulfur-containing materials may meet some of the requirements of the civil engineer require close consideration. Properties of sulfur-bonded concretes and mortars, sulfur composites, and sulfur–asphalt paving concretes indicate that in some applications, they are superior to conventional materials. Costs also are likely to be competitive, particularly in areas which are close to a source of supply. Present findings encourage the belief that with further work, materials incorporating sulfur will play an important role in the construction industry.

Acknowledgments

The authors wish to express thanks for conscientious technical help to Ron Begert, Harry Pollard, and Erhard Jeske and to colleagues in

UNISUL, especially J. B. Hyne (Chemistry) and E. J. Laishley (Biology) for helpful discussions. Financial support is acknowledged from the National Research Council of Canada through the auspices of UNISUL and from the Sulphur Development Institute of Canada.

Literature Cited

1. British Sulphur Corporation Ltd., London.
2. *Oilweek (Canada),* "Gas Processing Plant Capacities 1974" (1974) **24** (49), 30–32.
3. Pattison, D. A., "Oil Begins to Flow from Canadian Tar Sands," *Chem. Eng.* (1967) **74** (24), 66–70.
4. Stuart-Smith, J. H., Wennekers, J. H. N., "Bedded Sulphur in Canadian Arctic Islands," "Geology and Hydrocarbon Discoveries of Canadian Arctic Islands," *Am. Assoc. Pet. Geol. Bull.* (1977) **61** (1), 1–27.
5. *Miner. Yearb.,* "Metals, Minerals and Fuels" (1975) **1.**
6. Jimeson, R. M., Richardson, L. W., Needels, T. S., "Fossil Fuels and Their Environmental Impact," Symposium on Energy and Environmental Quality, Illinois Institute of Technology, Chicago, Illinois, 1974.
7. Southwest Research Institute, "Wall Coating," British Patent **1,405,961,** 1973.
8. Dale, J. A., Ludwig, A. C., "Sulphur Aggregate Concrete," *Civil Eng.,* *(ASCE)* (1976) **37** (12), 66–68.
9. Ludwig, A. C., "Demonstrating Structural Applications of Sulphur in Guatemala," *Sulphur Inst. J.* (Spring 1970) 14–17.
10. Crow, L. J., Bates, R. C., "Strength of Sulphur-Basalt Concretes," *U.S. Bu. Mines Rep. Invest.* (1970) **7349,** Spokane, Washington.
11. Ortega, A., Rybczynski, W., Ayand, S., Ali, W., Acheson, A., "The Ecol Operation," School of Architecture, McGill University, Montreal, Canada, 1972.
12. Malhotra, V. M., "Effect of Specimen Size on Compressive Strength of Sulphur Concrete," Mines Branch Report **74-25,** Dept. Energy Mines & Resources, Ottawa, Canada, June 1974.
13. Beaudoin, J. J., Sereda, P. J., "The Freeze-Thaw Durability of Sulphur Concrete," Nat. Res. Counc., Can., Div. Build. Res., Build. Res. Note **92** (June 1974).
14. Scrafford, E. C., "Cementitious Material," U.S. Patent **2,397,567,** April 1946.
15. McKinney, P. V., "Provisional Methods for Testing Sulfur Cements," *ASTM Bull.* (Oct. 1940) **96** (107), 27–30.
16. Brocchi, A., "Waterproof Cement," British Patent **2414,** Oct. 1860.
17. Peltic, W., "Materials for Filters, etc.," British Patent **1649,** 19 Dec. 1856.
18. Jordaan, I. J., Gillott, J. E., Loov, R. E., Hyne, J. B., "Effect of Hydrogen Sulphide on the Mechanical Strength of Sulphur and Sulphur Mortars and Concretes," *Mater. Sci. Eng.* (1976) **26** (1), 105–113.
19. Shrive, N. G., Gillot, J. E., Jordaan, I. J., Loov, R. E., "A Study of Durability in Temperature Cycles and Water Resistance of Sulphur Concretes and Mortars," *ASTM J. Test. Eval.* (Nov. 1977) in press.
20. Meyer, B., "Solid Allotropes of Sulphur," *Chem. Rev.* (1964) **64,** 429–451.
21. Meyer, B. (Editor), "Elemental Sulphur: Chemistry and Physics," Interscience, New York—London, 1968.
22. Wiewiorowski, T. K., Toure, F. J., "Molten Sulfur Chemistry. I. Chemical Equilibria in Pure Liquid Sulfur," *J. Phys. Chem.* (1966) **70,** 3528–3531.
23. Gillott, J. E., "Examination of Rock Surfaces with the Scanning Electron Microscope," *J. Microsc.* (1970) **91** (3), 203–205.

24. Shrive, N. G., Loov, R. E., Gillott, J. E., Jordaan, I. J., "Basic Properties of Some Sulphur-Bound Composite Materials," *Mater. Sci. Eng.* (1977) 30, 1, 71–79.
25. Dale, J. M., Ludwig, A. C., "Mechanical Properties of Sulfur," "Elemental Sulfur," B. Meyer, Ed., Chap. 8, Interscience, New York, 1965.
26. Rennie, W. J., Andreassen, B., Dunay, D., Hyne, J. B., "The Effects of Temperature and Added Hydrogen Sulphide on the Strength of Elemental Sulphur," Alberta Sulphur Research Quarterly Bulletin (1970) 7 (3), 47–60.
27. Davis, C. S., Hyne, J. B., "Thermomechanical Analysis of Elemental Sulphur: The Effects of Thermal History and Ageing," *Thermochim. Acta* (1976) 15 (3), 375–385.
28. Gamble, B. R., Gillott, J. E., Jordaan, I. J., Loov, R. E., Ward, M. A., "Civil Engineering Applications of Sulphur Based Materials," *Adv. Chem. Ser.* (1975) 140, 154–166.
29. "Durability of Concrete," *Am. Concr. Inst. Spec. Publ.* (1975) 47.
30. Gillot, J. E., "Alkali-Aggregate Reactions in Concrete," *Eng. Geol. Int. J.* (1975) 9, 303–326.
31. Payne, C. R., Duecker, W. W., "Chemical Resistance of Sulphur Cements," *Trans. Am. Inst. Chem. Eng.* (1940) 36 (1) 91–111.
32. Laishley, E. J., UNISUL, personal communications, 1975–1977.
33. Duecker, W. W., Estep, J. W., Mayberry, M. G., Schwab, J. W., "Studies of Properties of Sulfur Jointing Compounds," *J. Am. Water Works Assoc.* (1948) 49 (7), 715–728.
34. Carson, J. E., "Analysis of Soil and Air Temperatures by Fourier Techniques," *J. Geophys. Res.* (1968) 68 (8), 2217–2232.
35. Taber, S., "Frost Heaving," *J. Geol.* (1929) 37, 428–461.
36. Beskow, G., "Soil Freezing and Frost Heaving with Special Applications to Roads and Railroads," Swedish Geol. Soc., 26th Year Book, 3, Ser. C, 375, 1935.
37. Gifford, P., University of Calgary, unpublished data, 1976.
38. Bencowitz, I., Boe, E. S., "Effect of Sulphur Upon Some of the Properties of Asphalts," *ASTM Proc.* (1938) 38, Pt. II, 539–547.
39. Saylak, D., Gallaway, R. M., Epps, J. A., Ahmad, H., "Sulphur-Asphalt Mixtures Using Poorly Graded Sands," *Transport. Eng. J.* (1975) 101 (TE1), 97–113.
40. Lee, D. Y., "Modifications of Asphalt and Asphalt Paving Mixtures by Sulfur Additives," Iowa State University, ERI Project 834-5, Iowa, 1971.
41. Hammond, R., Deme, I., McManus, D., "The Use of Sand-Sulfur-Asphalt Mixes for Road-Base and Surface Applications," *Proc. Can. Tech. Asphalt Assoc.* (1971) 16, 27–52.
42. Kennepohl, G. J. A., Logan, A., Bean, D. C., "Conventional Paving Mixes with Sulfur-Asphalt Binders," *Proc. Assoc. Asphalt Paving Technol.* (1975) 44, 485–518.
43. Societé Nationale des Petroles d'Aquitaine, "New Bitumen- and Sulphur-Composition Based Binders and Their Process of Preparation," U.K. Patent 1303318, 1973.
44. Pronk, F. E., Soderberg, A. F., Frizzell, R. T., "Sulphur Modified Asphaltic Concrete," *Proc. Can. Tech. Asphalt Assoc.* (1975) 20, 135–194.
45. Sullivan, T. A., McBee, W. C., Rasmussen, K. L., *U.S. Bur. Mines Rep. Invest.* (1975) RI. 8087.
46. Shrive, N. G., Ward, M. A., "Some Low Temperature Ageing and Durability Aspects of Sulphur/Asphalt Concretes," in "Low Temperature Properties of Bituminous Materials," C. Marek, Ed., ASTM STP No. 628, 1977.

RECEIVED April 22, 1977. Authors are Civil Engineering members of the University of Calgary Interdisciplinary Sulfur Research Group (UNISUL).

Sulfur–Asphalt Binder Technology for Pavements

G. J. KENNEPOHL and L. J. MILLER

Gulf Oil Canada Limited, Research and Development Department, Sheridan Park, Ontario, Canada

A commercial process has been developed by Gulf Oil Canada to produce a novel sulfur–asphalt (SA) binder for flexible pavements by extending and replacing asphalt with up to 50 wt % elemental sulfur. The process uses a sulfur–asphalt mixer module (SAM) to finely disperse liquid sulfur into a continuous asphalt phase without the use of additives. The resulting binder has new and improved chemical and rheological properties resulting from interaction of sulfur with asphalt. Results from extensive laboratory evaluation such as improved flow characteristics (by Weissenberg rheogoniometer), improved temperature susceptibility, better fatigue (under stress-controlled testing), water resistance, and true material response as demonstrated in several test roads, are incorporated into comprehensive, fundamentally based pavement design technology.

While the significance of sulfur through its biological phase in nutrition, pharmaceutical, and fertilizer technology is well established, utilization of elemental sulfur as a construction material is still developing. Sulfur–asphalt binders for pavements appear a most significant development because of the potential for application and associated economic implication in highway construction and maintenance. A variety of approaches and methods to incorporate sulfur in pavements have been proposed, tested, and reported in the literature (*1–11*).

Based on resource management considerations which recognize future energy supply problems, dwindling hydrocarbons, and an increase in sulfur supplies because of pollution abatement, Gulf Oil Canada Ltd. has engaged in a major research and development program to develop a

0-8412-0391-1/78/33-165-113$10.00/0

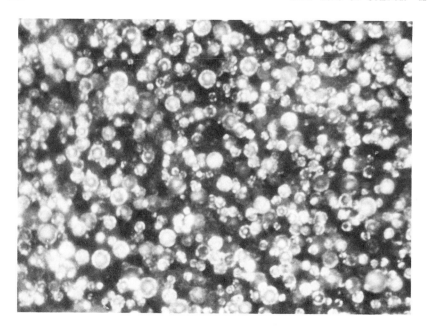

Figure 1. Photomicrograph of sulfur/asphalt dispersion (Magnification, 410×)

sulfur–asphalt (SA) process for conventional, flexible pavements. Various aspects of the Gulf Oil Canada SA process have been discussed in previous publications (11–16). The process involves the extension of asphalts by dispersion of liquid elemental sulfur into hot asphalt yielding a sulfur–asphalt (SA) binder. Up to 50 wt % of the asphalt can be substituted by sulfur, cutting down the effective asphalt content in the mix. The SA binder and SA binder-based mixes have been evaluated in extensive laboratory testing and also in several full-scale test roads in the provinces of Alberta, Saskatchewan, and Ontario.

Three tasks were pursued in the research and development phase of the Gulf Canada SA process: SA binder characteristics, processing technology, and design technology. This chapter summarizes and reports the progress which has been made towards commercial application of the acquired technology.

SA Binder Characteristics

SA Binder Concept. The principal objective of using elemental sulfur as a binder for aggregate without adversely affecting the flexibility of the in-place pavement guided the basic research approach and led to pre-mixing of sulfur and asphalt and to the SA binder concept first proposed by Bencowitz (3, 4). It is possible and tempting to produce by simple pugmill operations SA mixes with increased stiffness and stability as measured by the standard Marshall test methods. However, more sensitive testing indicates that not only the bulk material composition but also the sulfur–asphalt and the sulfur–aggregate interaction and microstructure affect uniformity, integrity, and performance of the SA pavement.

Sulfur–Asphalt Interaction. Sulfur and asphalt appear essentially immiscible and unreactive at the suggested mixing temperature of about 116°–160°C, the melting point and start of sulfur polymerization, respectively. As shown in the photomicrograph in Figure 1, high shear rate mixing produces a fine dispersion of liquid sulfur in asphalt with an average particle size of 4–5 μm. At room temperature the finely dispersed sulfur is crystalline. However, thermal analysis by differential scanning calorimetry (DSC) has shown that a substantial amount of elemental sulfur added to the asphalt has reacted or "dissolved" and is no longer present in crystalline form.

The relative amounts of crystalline and dissolved sulfur as determined by a Perkin Elmer DSC-1B is depicted in a block diagram, Figure 2. An investigation of ageing characteristics and stability of the non-crystalline portion of the SA binder with three grades of paving asphalts at four levels of sulfur concentration indicates that the amount of dis-

THERMAL (DSC) ANALYSIS

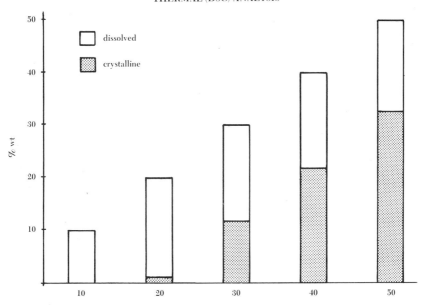

Sulfur added, % wt.

Figure 2.   Thermal (DSC) analysis

Table I. Aging Study

Sample Description	Wt % Elemental Sulfur Added	Total Wt % Sulfur in Binder
40/50 Pen	—	4.3
40/50 Pen	50	52.5
40/50 Pen	40	43.3
40/50 Pen	20	23.2
85/100 Pen	—	4.8
85/100 Pen	50	52.9
85/100 Pen	40	42.7
85/100 Pen	20	23.9
300/400 Pen	—	3.3
300/400 Pen	50	51.4
300/400 Pen	40	43.3
300/400 Pen	20	19.5

[a] Differential scanning colorimeter, Perkin–Elmer DSC-1B, was used to deter-

solved sulfur is independent of the asphalt grade and sulfur content and that no recrystallization of the dissolved sulfur occurs at ambient temperatures with time as shown in Table I.

The form of the dissolved sulfur has not been characterized properly yet. While stable at ambient temperatures, a substantial amount can be converted to crystalline sulfur at elevated temperatures or by solvent separation. This observation led to the development of a rapid liquid chromatography method to determine elemental sulfur in SA binders. The procedure which has been described previously by Cassidy (17) is based on gel permeation principle and uses a Styragel column and a uv detector. Results showed that 2–14% of the elemental sulfur added reacted chemically with the asphalt. Petrossi (18) and Lee (19), who determined free sulfur by extraction with sodium sulfite followed by titration with iodine, calculated a higher percent of bonded sulfur in sulfur–asphalt compositions. The observed differences are most likely caused by variations in the asphalt composition with regard to polar aromatics and naphthene components as well as by reaction temperature and contact time.

Electron paramagnetic resonance measurements have confirmed the existence of polysulfenyl radicals in liquid sulfur which competitively abstract hydrogen or which react by insertion or addition to give carbon sulfur bonds (20, 21). Hydrogen abstraction and subsequent hydrogen sulfide formation occurs predominantly above 160°C and must be avoided by reducing the mixing temperature. At temperatures below 160°C the prevalent reaction is believed to be the sulfur addition into the asphalt molecule (18, 19, 22, 23).

on Sulfur–Asphalt Binders

% Noncrystalline Sulfur in Total Binder[a] after Binder Preparation

2 Hr	8 Hr	2 Days	18 Days	46 Days	7 Mo	11 Mo
17.0	18.3	19.4	15.2	15.8	18.2	20.3
16.7	19.8	19.9	17.3	18.2	20.3	20.8
18.0	18.8	18.1	17.0	19.8	18.9	18.7
18.4	19.8	19.3	16.5	20.7	18.7	22.0
19.3	17.8	19.6	15.6	19.2	18.7	18.4
18.4	18.2	18.3	19.9	18.8	19.7	18.8
16.9	16.6	18.6	—	18.7	16.9	15.9
18.1	19.7	19.6	15.8	18.7	21.1	19.8
14.9	15.4	15.1	14.1	—	14.3	14.5

mine the quantity of crystalline sulfur present in the sulfur–asphalt binder.

Rheology of SA Binders. Conventional test methods such as softening point, viscosity, penetration, Fraas break point, ductilities, etc. have been used to characterize the rheology of SA binders (*11*). The physical structure of SA binders is complex, and the sulfur–asphalt and sulfur–aggregate interaction make correlations to asphalt and to binder properties for aggregate rather difficult.

While asphalt itself consists of a complex colloidal dispersion of resins and asphaltenes in oils, introduction of liquid elemental sulfur, which on cooling congeals into finely dispersed crystalline sulfur particles and in part reacts with the asphalt, necessarily complicates the rheology of such a SA binder. Differences and changes with SA binder preparation, curing time, temperature etc. must be expected and may be demonstrated by viscosity characteristics.

The viscosity of sulfur–asphalt binders has been determined over a wide temperature range, 4°–149°C. The Weissenberg rheogoniometer was used to obtain the absolute viscosity at temperatures between ambient and 120°C. Zeitfuchs cross arm (ASTM D2170) was also used to measure viscosity at 120°, 135°, and 149°C. Measurements made at 120°C with the Weissenberg rheogoniometer and Zeitfuchs cross arm are equivalent. Since the asphalt and sulfur–asphalt were Newtonian at higher temperatures, the cross arm was used to determine the viscosities reported at 120°, 135°, and 149°C. The Halekainen sliding plate microviscometer was used to determine the viscosity at 4°C.

The absolute viscosities (poises) for three grades of asphalt—40/50 pen, 85/100 pen, and 300/400 pen—at four levels of sulfur are plotted in Figures 3, 4, and 5. At lower temperatures where non-Newtonian

behavior was observed, the viscosity shown in Figures 3–5 is at a shear
rate of 5×10^{-2} sec^{-1}. As shown, the viscosity depends on temperature,
sulfur content, and asphalt grade. At higher temperatures the viscosity
for all binders containing dispersed sulfur drops below that of the same

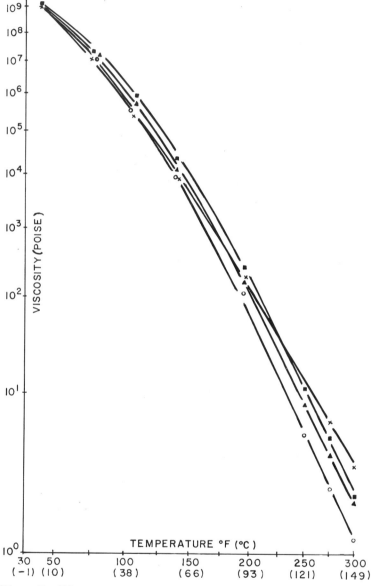

*Figure 3.   Viscosity vs. temperature at 5×10^{-2} sec^{-1}. ($\times$) $= 0/100$
SA–40/50 pen; ($\bigcirc$) $= 20/80$ SA–40/50 pen; ($\blacktriangle$) $= 40/60$ SA–40/50
pen; ($\blacksquare$) $= 50/50$ SA–40/50 pen.*

asphalt. The extent and temperature where this phenomenon occurs depends upon sulfur content. At lower temperatures sulfur–asphalt binders containing more than 20% sulfur have viscosities above that exhibited by the asphalt. This increased viscosity appears to be maximized

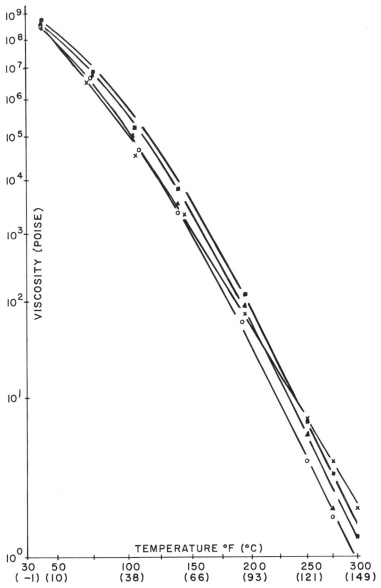

Figure 4.   Viscosity vs. temperature at 5 × 10⁻² sec⁻¹. (×) = 0/100 SA–85/100 pen; (○) = 20/80 SA–85/100 pen; (▲) = 40/60 SA–85/ 100 pen; (■) = 50/50 SA–85/100 pen.

in the temperature range, 25°–60°C. At 4°C the viscosity differential between the original asphalt and sulfur–asphalt has narrowed. The data indicates that the original asphalt is softened by the dissolved and liquid-dispersed sulfur. Upon cooling, however, the dispersed sulfur solidifies and increases the viscosity depending on the amount of crystalline sulfur.

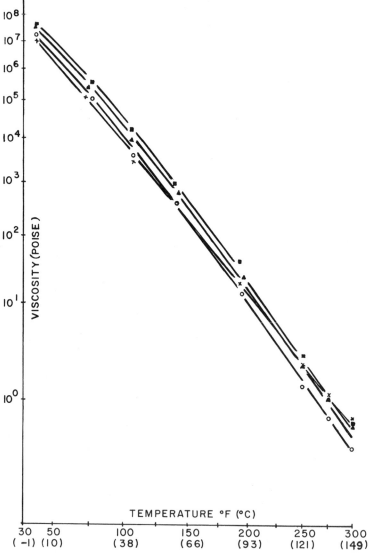

Figure 5.   Viscosity vs. temperature at 5×10^{-2} sec^{-1}. ($\times$) = 0/100 SA–300/400 pen; ($\bigcirc$) = 20/80 SA–300/400 pen; ($\blacktriangle$) = 40/60 SA–300/400 pen; ($\blacksquare$) = 50/50 SA–300/400 pen.

Figures 6 and 7 show viscosities of asphalt and sulfur–asphalt as a function of shear rate at 90° and 25°C. These plots show that the asphalts and sulfur–asphalts are Newtonian at 90°C. At 25°C the asphalts and sulfur–asphalt exhibit some non-Newtonian behavior. The degree of non-Newtonian behavior does not appear to be affected by sulfur concentration.

Sulfur–Aggregate Interaction. Substantial increases in stiffness of SA binder-based mixes, as measured by the Marshall test, tensile strength, and resilient modulus of elasticity, have been observed with increasing sulfur–asphalt ratio of the binder used (*11, 15, 16*). Such increases can, of course, be attributed in part to the increase in viscosity of the SA

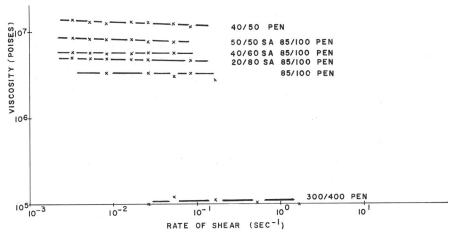

Figure 6.   Viscosity vs. rate of shear at 25°C (77°F)

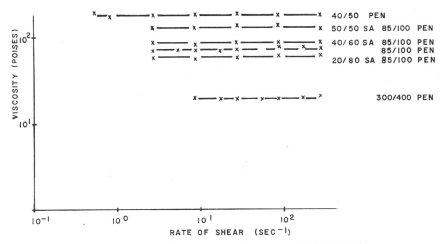

Figure 7.   Viscosity vs. rate of shear at 90°C (194°F)

binder. However, a sulfur–aggregate interaction could also account for high stiffness values of SA based mixes through increase in angularity of the aggregate, bridging or bonding of aggregate particles, and filling of voids.

Processing Technology

Premixing. The development and engineering of the SA processing technology is based on extensive laboratory and field experience with SA binder and SA mixes. The key aspects and requirements for the incorporation of sulfur in paving mixes for industrial-scale mix production are uniformity of dispersion, SA ratio, and coating of the aggregate by the binder.

While shearing in the pugmill can affect dispersion of the sulfur, it is still a function of the pugmill, the type and gradation of the aggregate, the binder content, mixing temperature, and mixing time. For instance an increase in the mixing cycle from 30 to 60 sec, which will curtail the throughput capacity of the pugmill considerably, has been reported (*24*). Although pugmill blending appears simple and easy, premixing of sulfur and asphalt, which requires only an additional in-line blending operation, is preferred because of the following five features: control of dispersion, control of SA ratio, uniformity of coating, reduced mixing cycle, and sulfur–asphalt contact and interaction. Just a comparison of costs for quality assurance with the small incremental cost for an in-line blender shows that premixing is the most economical route for commercial applications.

Sulfur–Asphalt Module. An integral part of SA binder hot-mix production is the sulfur–asphalt mixing module (SAM). In conventional mixing stations asphalt is metered into the pugmill where it is mixed with dried aggregate. The two components, aggregate and asphalt, are fed to the pugmill with appropriate feed control. SAM permits the addition of the third component, sulfur. The asphalt feed system can be by-passed and replaced by SAM which can deliver the desired SA binder or asphalt only. SAM is designed to produce SA binder at commercial rates up to 80 tons per hr—enough binder to produce 1000 tons per hr hot mix—and can be adapted to continuous and batch-operated mixing plants.

After two years experience with various prototypes, a commercial SAM unit has now been designed using only components which, with the exception of the mixer, are normally used in conventional mixing plants. This unit has been built by a leading supplier of equipment for pavement construction and has the following features:

(1) Binder levels are selected and then automatically integrated and adjusted to the operation of the pugmill by means of a computer.

(2) The sulfur–asphalt ratio and the binder level can be changed on the run.

(3) No modification to the mixing station is required other than addition of sulfur storage.

(4) The plant can always operate with asphalt only.

(5) No SA binder storage or surge tank is used.

(6) No more than 5–10 gal of SA binder (depending on length of product feed line) is present in the system, which can be flushed out with asphalt if so desired.

(7) Since components are common to the mixing plant, they have proven reliable and can be serviced by the regular crew.

Mix Production. The feed control system of the mix plant is integrated with the automatic controls on SAM. SAM automatically controls the sulfur–asphalt ratio and matches the production of SA binder to the aggregate feed rate. The SA binder is mixed with the aggregate between 116° and 160°C. This reduces the levels of emissions and produces mixes that are well coated and appear virtually indistinguishable from conventional asphalt mixes.

The SA binder is tested for dispersion and particle size prior to mix production with a microscope. The binder level of the mix is constantly measured with a Troxler model 2226 asphalt content gauge. Hot solvent extraction (ASTM D2172) using tetrachloroethylene solvent can also be used to measure the binder content of a SA mix. The sulfur–asphalt ratio of the binder is monitored in the field with the Troxler or by density measurements. Other methods that can be used to measure SA ratios are x-ray fluorescence of solutions of sulfur–asphalt in tetrachloroethylene, liquid chromatography, and differential scanning calorimetry. X-ray fluorescence measures total sulfur, liquid chromatography determines elemental sulfur, and DSC monitors crystalline sulfur.

Construction Procedure. Paving with a sulfur–asphalt hot mix is exactly the same as with asphalt. Conventional dump trucks are used to haul the hot mix from the asphalt plant to the road. There the mix is spread with standard pavement equipment and compacted with conventional rollers. Measurements on the completed pavements verify that sulfur asphalt mixes have good workability and are virtually indistinguishable from conventional asphalt.

Emissions. An environmental assessment has been carried out to evaluate the effects of sulfur–asphalt during and after construction. Ambient air samples were taken from points around the pugmill and paver and analyzed for hydrogen sulfide (H_2S), sulfur dioxide (SO_2), carbon disulfide (CS_2), carbonyl sulfide (COS), mercaptans (RSH), and total hydrocarbons. The results indicated that no problems exist in terms of current health standards during construction or after. Typical test results obtained at the pugmill and methods used are given in Table II.

Table II. Comparison of Emissions

Pollutant	Method[a]	Typical Measurements at Pugmill (ppm)	
		Asphalt	SA Binder
Hydrogen sulfide (H_2S)	Methylene-blue (American Public Health Assoc. method 701)	0.2	1.8[b]
Sulfur dioxide (SO_2)	West–Gaeke	0.1	0.2[b]
Carbon disulfide (CS_2), carbon sulfide (COS), and mercaptans (RSH)	Potentiometric titration with silver nitrate UOP method 163-67	< 0.1	< 0.1
Elemental sulfur	Particulate filter total sulfur	—	0.5
Total hydrocarbons	Beckman model 400 hydrocarbon monitor	6	6

[a] Methods acceptable to Environmental Protection Agency (EPA) or Environment Canada.
[b] American Governmental Conference on Toxicology and Health (AGCTH) set threshold limits for H_2S at 10 ppm and for SO_2 at 5 ppm.

Soil and water quality measurements before and after road construction have indicated no change in sulfur content and pH.

Test Roads. Three test roads were built in the provinces of Alberta and Ontario in 1974 and 1975. Two more full-scale demonstration projects were carried out in Ontario and Saskatchewan in 1974 and 1976. Table III summarizes the field tests.

The field trials have demonstrated the feasibility of full-scale mix production and use of conventional equipment for transporting, laying, and compacting. Another test road is scheduled for June 1977 in Michigan in cooperation with the Michigan Department of State Highways and Transportation.

Performance. All test roads are monitored and evaluated with the cooperation of the Alberta Research Council and the Highway Departments of Alberta and Ontario. The in-service evaluations include the following periodic measurements of structural behavior and performance.

(1) Rebound tests (Benkelman beams)
(2) Surface wave tests and dynamic moduli
(3) Temperature of subgrade and pavement
(4) Surface elevations changes
(5) Density by road logger

(6) Photographic crack mapping

(7) Roughness

Results from continuing in-service evaluation of all test roads are being collected and used to develop comprehensive design and application recommendations as well as correlations with analytical and theoretical laboratory analyses.

Design Technology

Engineering Evaluation. The use of elemental sulfur in paving mixes provides new engineering properties and also gives an interesting new dimension to flexible pavement design. A comprehensive engineering evaluation has been initiated not only to study all aspects of the sulfur–asphalt process but also to identify and quantify the significant factors required to develop a design technology for SA-based paving mixes and their application. The flow chart in Figure 8 indicates the relationship of the engineering evaluation to the development of the process and the field tests and the involvement of closely coordinated field and laboratory testing.

Table III.   Summary of Field Tests

	1974, Port Colborne, Ontario	1974, Blue Ridge, Alberta	1974, Windfall, Alberta	1975, Renfrew, Ontario	1976, Melville, Saskatchewan
Mix type, top size (in.)	3/8	5/8	3/4	5/8 sand	3/4
Asphalt (pen.)	85–100	300–400	300–400	150–200	AC 1.5
Sulfur (wt %)	50	40, 50	40, 50	40, 50	20, 35, 50
Construction					
overlay	X		X		X
full depth		X		X	
Mixing station					
batch	X			X	
continuous		X	X		
drum drier					X
Paver					
conventional	X	X	X	X	X
grader			X		
Roller					
steel	X	X	X	X	
vibratory		X	X		X
rubber tired		X	X	X	

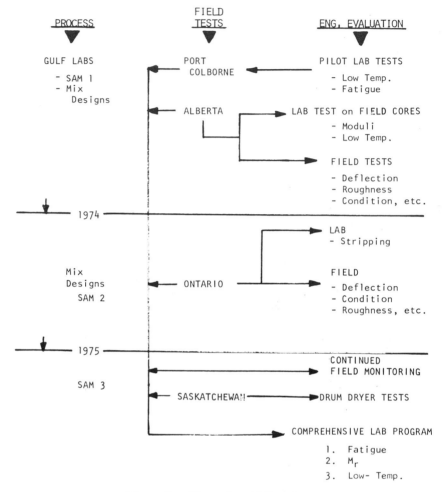

Figure 8.   Chronological flow chart

The major variables identified and investigated using laboratory and field test results include:

(1) SA ratio
(2) Binder content
(3) Temperature
(4) Asphalt grade
(5) Asphalt temperature susceptibility
(6) Stress level (fatigue)
(7) Layer thickness (full-depth test road)

Various types of engineering evaluation, both field and laboratory, are required to establish engineering properties of the mixes made with the new SA binder in order to permit an assessment of structural response (i.e., fatigue, permanent deformation, and shrinkage fracture). With such knowledge of the engineering properties, it will be possible to make comparisons with standard conventional paving mixes and eventually to conduct fundamental structural (layer) analyses. Such investigation will lead to methods for predicting the overall performance or service lives for SA-based pavements. The verification of such methods depends upon continuing in-service evaluation of existing SA test roads.

The continuing or periodic field evaluations on the Alberta SA test roads are being conducted by the Alberta Research Council and Alberta Transportation while the Ministry of Transportation and Communication of Ontario is monitoring the Ontario test sections. Several years of measurements and observations are needed in order to draw conclusions from field evaluations.

A comprehensive laboratory testing program is being conducted in addition to compiling and evaluating test data from the various field trials. The program involves the application of various analytical methods of structural analyses to evaluate engineering properties such as tensile strength, resilient modulus of elasticity, fatigue life, low-temperature characteristics, durability, and permanent deformation. Completed to date are two separate studies on low-temperature characteristics and fatigue carried out at the University of Waterloo, Canada and by Austin Research Engineers, Texas. The procedure and results have been published in some detail (*15, 16*). The low-temperature evaluation and the fatigue evaluation both involve factorial designed experiments in order to identify any potential interaction of the variables.

For the low-temperature experiment the rate of extension apparatus, as described by Hajek (*25*) and Haas (*26*), was used to determine the stiffness of test specimen of SA binder based concrete mixes at $-10°$ and $-25°C$. The asphalt cements chosen were 40–50, 85–100, and 300–400 penetration grade to cover the entire range as well as to ensure that if a difference in properties did exist, it would be noticeable. The other test variables included temperature susceptibility (obtained by using asphalts based on crude oils from various sources), SA ratio, and binder content.

The tensile strength and the resilient modulus of elasticity were measured as part of the fatigue life evaluation. The basic test method was the static and repeated load indirect tensile test which involves loading a cylindrical specimen with a single or repeated compressive load. The loading develops a relatively uniform tensile stress perpendicular to the direction of the applied load and along the vertical diametral plane which ultimately causes the specimen to fail by splitting along the vertical diameter.

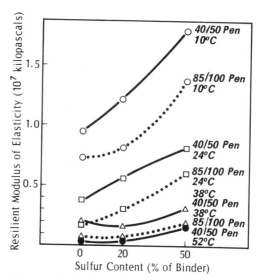

Figure 9. Relationships between resilient modulus and sulfur content

In the repeated load tests, the specimens were subjected to a 1.0-Hz load pulse in which the load was applied for 0.1 sec and was followed by a rest period of 0.9 sec. During testing, the resulting horizontal deformations were recorded using a light-beam oscillograph. The sequence of testing was to establish the tensile strength and then to use this information to generate fatigue life vs. stress relationship under dynamic loading, where the stress is chosen as a percentage of the strength. Resilient modulus of elasticity was measured as part of the dynamic testing for fatigue life.

Major Findings and Implications. The data from low- and high-temperature factorial designed experiments were statistically analyzed. The conclusions are summarized below.

Low-Temperature Stiffness. Results from the low-temperature experiment indicated that the stiffness was significantly affected only by asphalt penetration. Stiffness is directly related to low-temperature cracking, i.e., as stiffness increases the potential for cracking or the number of cracks increases (25, 26). Adding sulfur (SA ratio), changing the binder content, or varying the viscosity did not appear to affect the stiffness significantly over the range investigated. It is, therefore, the nature of the asphalt itself in a SA binder which controls the low-temperature cracking potential of the binder. Thus in no way has the addition of sulfur affected adversely the low-temperature characteristics of the asphalt concrete.

TENSILE STRENGTH. The strength of the SA mixes was higher than the strength of the conventional mixes. Strength values ranged from 10.3 to 465 N/cm and were found by analysis of variance to be affected significantly by the SA ratio, penetration of the asphalt cement, and temperature. The addition of sulfur in excess of 20% increased the strength significantly. The increase in strength produced by adding 50% sulfur was approximately equal to the increase associated with using a 40/50 penetration asphalt cement rather than an 85/100 penetration.

RESILIENT MODULUS OF ELASTICITY. The addition of sulfur produced an increase in stiffness in terms of resilient modulus of elasticity, Figure 9. Generally this increase required the addition of 50% sulfur, although in some cases 20% sulfur produced a substantial increase.

FATIGUE LIFE (STRESS-CONTROLLED TESTING). The addition of sulfur in excess of 20% produced a significant increase in fatigue life. There was some indication that a much greater increase in fatigue life could be expected at this level of sulfur where the high-temperature susceptibility asphalt was used. The increase in fatigue life produced by adding 50% sulfur was approximately equal to the increase associated with using a 40/50 pen asphalt rather than an 85/100 penetration.

The results from laboratory engineering evaluation and field verification have interesting design implications in that the use of elemental sulfur provides the designer with an extra degree of flexibility. The grade of asphalt and the SA ratio can be revised to gain desired mix characteristics at low and at high service temperature.

The design implications can be demonstrated best by the concept of an ideal mix from Hignell et al. (27) which is illustrated in Figure 10. The bottom dashed line gives the characteristics of a mix optimized for

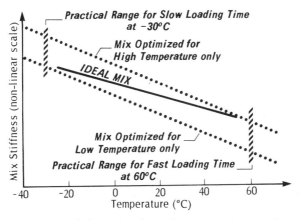

Figure 10.   Schematic of consistency vs. temperature requirements for asphalt paving mixtures

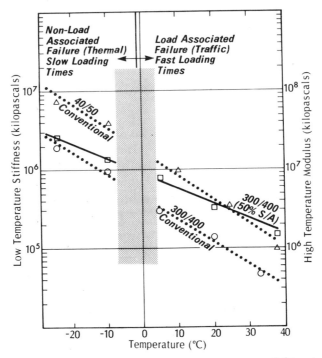

Figure 11. Modification of temperature susceptibility of asphalt mixtures with sulfur

low-temperature application with a stiffness low enough to resist low-temperature cracking but with an insufficient stability at high service temperature. The top dashed line shows a mix optimized for high-temperature application while the solid line represents the characteristics of a mix which is optimum for both low- and high-temperature conditions.

Figure 11 illustrates that SA mixes tend to approach the ideal mix. The top line in the figure represents a conventional mix with 40/50 penetration asphalt, the bottom line a conventional mix with 300/400 penetration asphalt, and the solid line a SA mix with 300/400 penetration asphalt containing 50 wt % sulfur. The results indicate that at low temperature, the 300/400 SA mix has properties similar to the conventional 300/400 mix, and at high temperatures the 300/400 SA mix has a stiffness in the range of that for conventional 40/50 mix. Thus the 300/400 SA mix will crack less than a conventional 40/50 mix at low temperatures and will have high strength and stability at high temperatures.

The type of plot shown in Figure 11 combines low-temperature stiffness modulus at slow loading (thermal) with high temperature stiffness at fast loading times (traffic, using a decade shift in the values of the left and right ordinates. A transition zone of about −6° to +4°C has been used to distinguish between the two situations. While each side of the diagram refers to a different type of loading situation, and projection of the lines across the whole diagram may be questioned, it is a very useful way to illustrate mix-temperature susceptibility over the entire service temperature range. In addition, as subsequently illustrated, it proved applicable to the entire range of mixes and variables of the investigation.

Economic Aspects

An evaluation of economic aspects of SA pavement systems involves, in addition to an assessment of various performance benefits, the relative costs for equipment, labor, materials, construction requirements, etc. Performance advantages and cost savings that give rise to significant economic benefits are outlined in subsequent paragraphs.

Improved Performance. A key aspect in SA-binder-based design technology is that sulfur provides an additional variable for flexible pavement design. The Gulf Canada SA process uses advantageously SA ratio, binder content type of asphalt, and aggregate as design variables to improve and to optimize performance. Results from extensive engineering evaluation have shown substantially higher performance of SA mixes with respect to low-temperature applications (*16*), increased tensile strength, and fatigue life under stress-controlled testing (*15*).

Resistance to Water. A common cause of pavement failure in many areas is stripping of the asphalt from the aggregate by the action of water. The tendencies of stripping by water can be reduced greatly by

Table IV. Resistance to Stripping by Water

By Modified ASTM D-1664

% Binder Coating Remaining on Aggregates

Total Immersion in Water (min)	150/200 Pen	40:60 Sulfur/Asphalt	50:50 Sulfur/Asphalt
15	80	90	95
30	75	90	95
45	75	85	95
60	70	80	90
240	55	80	90

By Modified ASTM D-1075

% Retained Marshall Stability of Binder

Mix Type	85/100 Pen	40:60 Sulfur/Asphalt	50:50 Sulfur/Asphalt
ASTM 5A	80	100	100
Sand	70	100	100

incorporating sulfur in the paving mix. Results of a modified method of test for "Coating and Stripping of Bitumen Aggregate Mixtures" (ASTM D1664) and the standard method of test for "Effect of Cohesion of Compacted Bituminous Mixtures" (ASTM D1075) are given in Table IV. The data show that the retention of a bituminous film by a quartz aggregate submerged in water at 60°C is considerably improved and that the compressive strength of Marshall briquettes submerged in water at 25°C for seven days is retained fully.

Resistance to Fuel. Damage of asphalt concrete pavements by hydrocarbon fuels such as gasolines, jet fuels, and motor oil often occurs in particular on air terminals, parking lots, and driveways. It has now been found that SA-binder-based pavements have an improved resistance to fuel spillage. The compressive strength of Marshall briquettes made with SA binder was determined before and after contact with a jet fuel. The retained Marshall stabilities, as shown in Table V, are considerably higher for test specimens containing sulfur, indicating improved resistance to fuel.

Use of Low-Quality Aggregate. Good-quality aggregates are costly and in short supply in certain areas. Laboratory investigations have shown that some low-quality aggregates can be improved greatly especially with respect to Marshall stability. For example, the Marshall stability (ASTM D1559) of 608 for a sand mix, as shown in Table VI, is increased to 2608 when an SA binder is used.

Conventional Equipment. Besides a storage tank for sulfur and SAM for SA binder production, the Gulf Canada SA process requires no other modifications to the mixing station and paving equipment. The paving and construction requirements are entirely conventional. Field trials (*11, 12*) have shown that the workability of the SA-binder-based hot mix is good and that conventional equipment can be used for transporting, placing, and compacting.

Energy Saving. Reduced fuel requirements for heating the aggregate can be realized by decreasing the hot-mix temperature (estimated

Table V. Resistance of Sulfur–Asphalt Pavements to Fuel Spillage

Marshall Stability (lb)

Mix Design—ASTM 5A	Fuel	Origi- nal	Drip 6 Ml	Total Immer- sion 4 Hr	Total Immer- sion 4 Hr + 3 Days Drying
6%–85/100 pen asphalt (by weight)	jet 'A'	2858	2450	650	1590
6%–50:50 sulfur asphalt (by weight)	jet 'A'	5963	4650	3910	4750

Table VI.   Marshall Data for Sand Mix

Binder Type	Wt %	Voids (%)	Flow (0.01 in.)	VMA (%)	Stability (lb)
Asphalt[a]	9.0	5.7	9.5	25.1	608
40/60 SA	11.5	4.8	10.0	25.3	1210
50/50 SA	12.1	6.9	10.0	27.5	2608

[a] 85–100 pen.

10% for 30°C) because the viscosities of SA binders are below those of the original asphalts at higher (pugmill) temperatures as shown in Figures 3–5. Furthermore, the replacement of asphalt by sulfur will extend the asphalt supply and contribute to hydrocarbon energy conservation.

Conclusion

The development of the Gulf Oil Canada SA process for pavements approaches commercial application. The use of elemental sulfur in paving mixes gives new engineering properties and also adds an interesting new dimension to flexible pavement design. The key advantages of the process and major findings from laboratory and field tests are listed below.

(1) A significant portion of the asphalt can be replaced by elemental sulfur to yield conventional high-performance flexible pavements.

(2) Conventional equipment and methods can be used for mix production, material handling, and paving procedure.

(3) Clean and safe operations are maintained with respect to potential sulfur-based pollutants.

(4) The utilization of elemental sulfur in pavements with demonstrated simplicity of construction requirements is economically attractive.

(5) Low-quality aggregates and sands can be used.

(6) Softer asphalts may be used to reduce low-temperature cracking without the high-temperature deformation which occurs with the use of asphalt alone.

(7) SA mixes exhibit significantly higher fatigue lives than comparable conventional mixes under stress-controlled testing.

(8) The pavements have greater resistance to water.

Literature Cited

1. Abraham, H., "Asphalt and Allied Substances," 6th ed., Vol. II, p. 178, D. Van Nostrand Co., Inc., New York, 1961.
2. Duecker, W. W., *Proc. Natl. Paving Brick Assoc.* (1937) 60.
3. Bencowitz, I., et al., U.S. Pat. 2,182,837 (Feb. 1936).
4. Bencowitz, I., et al., ASTM Preprint (1938) 97 (9) 539.
5. Litehiser, R. R., Schofield, H. Z., *Proc. Highway Res. Bd.* (1936) 182.

6. Ariano, R., *Strade* (1941) 119.
7. Fike, H. L., "Sulfur Research Trends," ADV. CHEM. SER. (1972) **110**, 298.
8. Burgess, R. A., Deme, I., "New Uses of Sulfur," *Adv. Chem. Ser.* (1974) **140**, 85.
9. Garrigues, C., Vincent, P., "New Uses of Sulfur," ADV. CHEM. SER. (1974) **140**, 130.
10. Pronk, F. E., Soderberg, A. F., Frissell, R. T., *Can. Tech. Asphalt Assoc.* (1975) **20**.
11. Kennepohl, G. J. A., Logan, A., Bean, D. C., *Can. Sulfur Symp. (Pap.)* (May 1974).
12. Kennepohl, G. J. A., Logan, A., Bean, D. C., *Assoc. Asphalt Paving Technol.* (1975) **44**.
13. Kennepohl, G. J. A., *Energy Process Can.* 24 (July–Aug 1976).
14. Kennepohl, G. J. A., Proceedings of the Symposium on New Uses for Sulfur and Pyrites, Madrid, Spain, The Sulphur Institute, London, 1976.
15. Kennedy, T. W., Haas, R. C. G., Smith, P., Kennepohl, G. J. A., Hignell, E. T., *Trans. Res. Bd. Record* (1977), in press.
16. Meyer, F. R. P., Hignell, E. T., Kennepohl, G. J. A., Haas, R. J., *Assoc. Asphalt Paving Technol.* (1977) **46**.
17. Cassidy, R. M., *J. Chromatogr.* (1976) **117**, 71.
18. Petrossi, U., et al., *Ind. Eng. Chem. Prod. Res. Dev.* (1972) **11** (2) 214.
19. Lee, Dah-Yinn, *Ind. Eng. Chem. Prod. Res. Dev.* (1975) **14** (3) 171.
20. Fairbrother, F., et al., *J. Polym. Sci.* (1955) **14**, 459.
21. Gardner, D. M., *J. Am. Chem. Soc.* (1956) **78**, 3279.
22. Van Ufford, J. J. Q., et al., *Brennst.-Chemie* (1962) **43** (6), 173.
23. *Ibid.* (1965) **45** (1) 1.
24. Deme, I., *Assoc. Asphalt Paving Technol.* (1974) **43**.
25. Hajek, J. J., Haas, R. C. G., *Can. Technol. Asphalt Assoc.* (1971) **16**.
26. Haas, R. C. G., *Asphalt Inst., Rep.* (Jan. 1973) **RR 73-1**.
27. Hignell, E. T., et al., *Assoc. Asphalt Paving Technol.* (1972) **41**.

RECEIVED April 22, 1977.

Sulfur Utilization in Asphalt Paving Materials

WILLIAM C. McBEE and THOMAS A. SULLIVAN

U.S. Department of the Interior, Bureau of Mines, Boulder City Metallurgy
Engineering Laboratory, Boulder City, NV 89005

*Research is being conducted on potentially large-scale uses
of sulfur to extend or replace such construction materials
as asphalt, cement, and mineral aggregates. Paving materials have been developed in which sulfur replaces up to
50% of the asphalt and which can be handled with existing
mixing and paving equipment. Studies involving the use of
sulfur–asphalt mixtures as binders for good graded aggregates showed that the resulting materials were equal or
superior to those in conventional bituminous pavement.
Laboratory mix design studies and tests on mixtures containing poor-quality aggregates or graded sands demonstrated
that sulfur in the binder upgraded the properties of the
substandard aggregate and resulted in materials suitable
for normal paving uses.*

R ecently, interest in sulfur utilization in asphalt paving materials has
been rekindled. One of the foremost reasons for this is the potential
availability of surplus sulfur recovered from secondary sources in connection with meeting environmental pollution standards (*1*). These
sources include sulfur recovered from sour gas, from refining of petroleum,
and from smelter and powerplant stack gases. Another potential saving
of energy and petroleum is possible by replacing part of the asphalt
binder with sulfur (*2*).

Sulfur as a replacement for aggregate in preparing sand–sulfur–
asphalt paving has been reported by Burgess (*3*), by Sullivan (*4*), and
by Saylak (*5*). This chapter emphasizes the substitution of sulfur for
part of the asphalt binder to produce sulfur–asphalt pavements with
properties equivalent or superior to those of conventional asphaltic
concrete paving. Several methods of doing this have been reported.
Bencowitz (*6*), in 1938, described a method of preparing the sulfur–

0-8412-0391-1/78/33-165-135$10.00/0

asphalt binder by mixing the two together with a stirrer before mixing with the aggregate. Garrigues (7) and Kennepohl (8) used an emulsified mixture of sulfur in asphalt as the aggregate binder. Test roads using this method have been placed in France, in Canada, and recently in the U.S. (9).

The Bureau of Mines program was directed toward developing a simplified method to prepare and utilize the sulfur–asphalt binder (10). Sulfur has a high solubility (20% or greater) in most asphalts at the mixing temperature range of 265°–300°F. This fact and the high shear energy developed in mixing the thin film of binder over the hot aggregate suggested that sulfur and asphalt could be introduced directly into the mixer and a paving mixture prepared without emulsifying the sulfur into the asphalt.

A test program to determine the feasibility of direct substitution of sulfur for asphalt in preparing sulfur–asphalt concretes was conducted. Properties of the resultant materials were compared with those of emulsified sulfur–asphalt binder materials and conventional materials. In addition, a test program was conducted to determine if direct-substituted binders could be used to upgrade marginal aggregates for use in paving materials.

Laboratory Evaluation Program

This program was divided into two sections. The first consisted of the preparation and evaluation of paving materials made with two types of standard aggregates using both direct-substituted and emulsified sulfur–asphalt binders with normal asphalt binders as a control. This evaluation primarily determined the Marshall properties of the different materials. In addition, a temperature sensitivity study was made on all materials using samples prepared under normal mixing procedures but compacted at temperatures ranging from 185° to 305°F. Five levels of sulfur replacement were tested. The Marshall properties of the temperature sensitivity samples were determined and compared. Fatigue tests at constant stress and dynamic stiffness measurements were also performed on the materials prepared by the direct-substitution method.

The second part of the program was devoted to upgrading marginal aggregate materials using a direct-substituted sulfur–asphalt binder. Three types of sand aggregates and a waste limestone chat were evaluated. The sands were graded to meet uniform specifications developed by the Texas Highway Department for hot sand–asphalt paving mixtures. The limestone chat was a waste material from a limestone quarry. These materials were mixed with asphalt alone, with the five levels of direct-substituted sulfur–asphalt binders, and the products were evaluated for

Marshall properties and dynamic stiffness. Temperature sensitivity studies also were performed on these materials.

Experimental. MATERIALS. The sulfur was commercial grade (99.9% purity) secondary elemental sulfur and was furnished in flake form. Asphalt mixtures were prepared using a West Coast (Los Angeles, CA) viscosity-graded AR 2000 asphalt (ASTM D 3381-75) (*11*). The graded aggregates consisted of crushed limestone and volcanic rock, each blended with 25 wt % construction sand and 15 wt % desert sand to conform to Asphalt Institute specifications for a type IVb aggregate (*12*). A reject limestone chat material from the U.S. Lime Co. quarry at Apex, NV was tested as a marginal-grade coarse aggregate material. The aggregate gradations of these three materials are plotted in Figure 1 and compared with Asphalt Institute specifications.

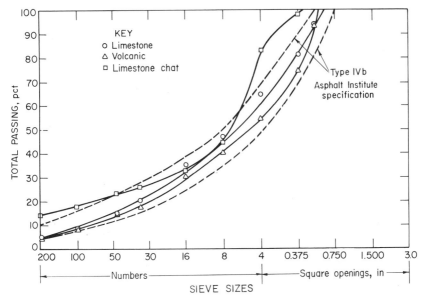

Figure 1.   Aggregate gradation for coarse aggregate materials

Substandard sand aggregates were investigated as marginal aggregate materials that could be upgraded using the higher stabilities afforded by sulfur–asphalt binders. The sands were a mixture of desert blow sand and 7 wt % minus 200-mesh silica flour; a blend of equal weights of volcanic mortar sand and mortar slimes; and a minus ⅛-in. limestone sand. The physical properties of these aggregate minerals are shown in Table I. Aggregate gradations for the sand materials are shown in Figure 2.

PREPARATION OF PAVING MATERIALS. Conventional asphaltic concrete is prepared with sufficient asphalt to coat the aggregate and reduce

Table I.   Physical Properties of Aggregates

Aggregate Type	Specific Gravity (bulk)[a]	Voids in Mineral Aggregates (%)[b]	Unit Wt (lb/ft³)
Type IVb, limestone	2.70	23.0	124.1
Type IVb, volcanic	2.59	24.9	121.4
Limestone chat	2.70	23.4	128.3
Limestone sand	2.69	29.2	110.2
Desert sand, 7% silica flour	2.68	30.3	112.2
Motor sand, 50% slimes	2.49	31.6	106.8

[a] Determined using ASTM method C127-73.
[b] Determined using ASTM method C30-37.

the compacted void content to approximately 3%. The amount of asphalt needed for a specific aggregate can be determined by the Marshall method of hot mix design (ASTM D1559-75). This method was used with all the aggregate blends to determine the optimum volume of asphalt required.

Sulfur was substituted for part of the asphalt required for normal graded aggregates using equal volumes of the binder to maintain a normal void level in the products. Since sulfur is twice as dense as asphalt, it requires approximately double the weight of sulfur to replace an equal volume of asphalt. Two methods of using sulfur in the binder were

Table II.   Marshall

Composition[a]				Specific Gravity[b]	
Sulfur		Asphalt			
vol %	wt %	vol %	wt %	D	E
Limestone Aggregate					
0	0	100	5.7	2.411	2.416
15	1.8	85	5.1	2.427	2.416
25	2.9	75	4.5	2.425	2.429
35	4.1	65	3.9	2.423	2.440
50	5.8	50	2.9	2.437	2.452
75	8.6	25	1.5	2.433	2.483
Volcanic Aggregate					
0	0	100	7.0	2.288	2.263
15	2.0	85	5.9	2.300	2.286
25	3.4	75	5.2	2.307	2.303
35	4.7	65	4.4	2.321	2.314
50	6.7	50	3.4	2.320	2.313
75	9.9	25	1.7	2.343	2.330

[a] Vol %, volume–percent in the binder; wt %, weight–percent of the total mix.
[b] D, direct-substituted sulfur mixes; E, mixes using sulfur–asphalt emulsion

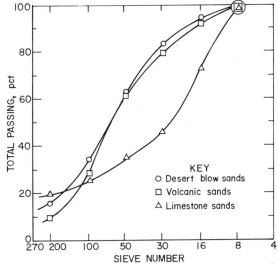

Figure 2. Aggregate gradation for sand materials

evaluated. They were a sulfur–asphalt emulsification method and a sulfur–asphalt direct substitution method.

Emulsification Method. Emulsified sulfur–asphalt binder was prepared by combining molten sulfur and liquid asphalt in a 2½-in.-diameter, vertical, Gifford Wood colloid mill (*1*) at a rotor-stator gap setting of

Design Data Comparison

Voids (%)[b]		Marshall				Dynamic Stiffness (psi × 10⁶)[c]
		Stability (lb)		Flow (0.01 in.)		
D	E	D	E	D	E	D
		Limestone Aggregate				
2.3	2.0	2,140	2,310	10	8	0.599
1.8	2.3	2,325	1,875	10	9	1.466
2.5	2.3	2,530	2,495	10	10	1.639
3.2	2.5	3,750	4,260	8	9	2.364
3.5	2.9	7,350	7,250	8	8	—
4.9	3.0	10,615	10,000	8	8	—
		Volcanic Aggregate				
2.4	3.3	2,580	2,430	12	11	0.653
2.8	3.2	2,230	2,575	10	11	0.980
3.2	3.3	3,085	3,000	11	11	1.542
3.2	3.6	5,520	5,055	10	9	2.040
4.4	4.5	9,605	9,540	12	11	—
4.9	5.4	9,910	10,000	6	8	—

binders.
[c] Determined using the Schmidt method (*13*) at ambient temperature.

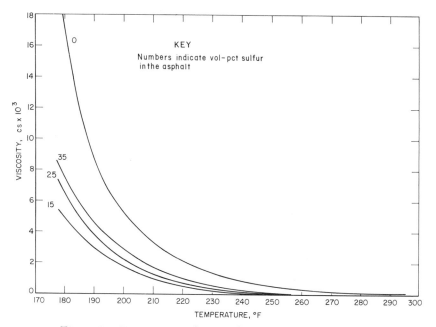

Figure 3. *Viscosity of sulfur–asphalt binder vs. temperature*

0.018 in. at 7000 rpm. The mix was recirculated for 8 min at 285°–305°F. The emulsified binder was immediately combined with heated aggregate (300°–320°F) and mixed in a laboratory-sized Hobart mixer for 2 min to prepare a sulfur–asphalt concrete at 300°F. Marshall samples were prepared by placing the mixture in Marshall molds and compacting with 75 blows of a 10-lb compaction hammer on each side of the sample at 270°F.

Direct Substitution Method. Sulfur and asphalt were poured together at 300°F and were added to the heated aggregate at 300°–320°F. The three materials were mixed in the Hobart mixer for 2 min. Standard Marshall test samples for Marshall and dynamic stiffness tests were prepared as in the emulsification method. Bar samples for fatigue testing were prepared by kneading the mix into $3 \times 3 \times 12$-in. bars to obtain compaction equal to the Marshall compacts.

Sand–Sulfur–Asphalt Method. The direct substitution method was used to evaluate the substitution of sulfur for asphalt in preparing paving materials with improved properties. Sulfur and asphalt were combined and added to graded sand in the Hobart mixer and were mixed for 2 min. Test samples were prepared as described for the direct substitution method.

Temperature Sensitivity Samples. Materials for sensitivity tests were mixed by the methods described above. Marshall specimens were compacted by specific temperatures in the range of 185° to 305°F. After mixing, the paving material was placed in compaction molds, cooled to the desired temperature, spaded with a small trowel 15 times around the outer perimeter and 10 times in the center, and then compacted with 75 blows of the 10-lb hammer on each side.

Results of Laboratory Testing. The optimum amounts of asphaltic binder for the standard type IVb limestone and volcanic aggregates were established as 5.7 and 7.0 wt %, respectively. Test samples were prepared for each aggregate by both the emulsified and direct substitution method using the asphalt binder alone and by substituting 15-, 25-, 35-, 50-, and 75-vol % sulfur for the asphalt. Properties obtained on testing these paving materials are shown in Table II. The results shown are the average of four separate determinations. Increasing the amount of sulfur

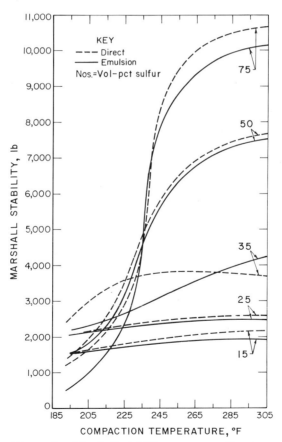

Figure 4.   Marshall stability vs. compaction temperature—limestone aggregate

in the binder results in increased stability and density of the resultant materials. There was little variation in the amount of flow in any of the samples. In general, there was no evidence of any differences in the products prepared by either method at a given sulfur substitution level.

Viscosity (Brookfield) changes resulting from substituting up to 35 vol % sulfur in the binder were determined, and the results are shown in Figure 3 along with the values for asphalt. The results show that sulfur lowers the viscosity of the binder at normal working temperatures and therefore should increase the efficiency of mixing and placing the concrete. In contrast to the viscosity lowering at elevated temperatures, the sulfur acts as a filler or structuring agent in the binder of the product when cooled to ambient temperatures. This was shown by the increasing stability with increasing sulfur content of the binder.

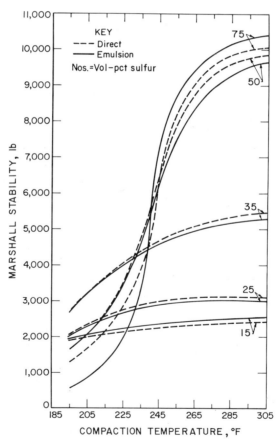

Figure 5.   Marshall stability vs. compaction temperature—volcanic aggregate

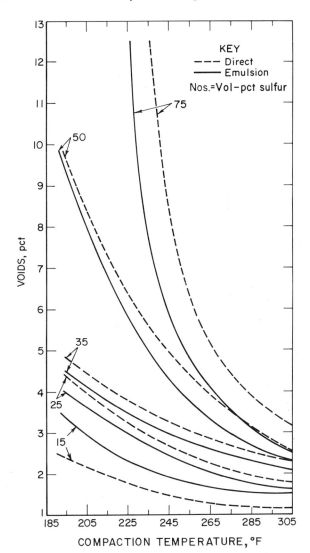

Figure 6. Voids content vs. compaction temperature—limestone aggregate

A temperature sensitivity study was made to establish the workability
of the paving materials with increasing sulfur contents in the binders.
Marshall compaction samples were prepared in quadruplicate for all the
mixtures shown in Table II, and the effect of compaction at 10 temperatures was determined. Figures 4 and 5 show the results of the temperature
sensitivity study on the Marshall stability of both methods of mixing with
each aggregate. Figures 6 and 7 give the void content of the same

materials prepared at various temperatures. Since the values obtained on testing samples prepared with no sulfur additions were similar to those obtained using a 15-wt % sulfur substitution, these values were omitted from the figures. There were no apparent major differences in properties prepared by either method. From the results, it appears that sulfur–asphalt paving materials containing up to 35 vol % (52 wt %) sulfur in the binder can be prepared and placed on roadways with essentially the same mixing plants and paving equipment and at the

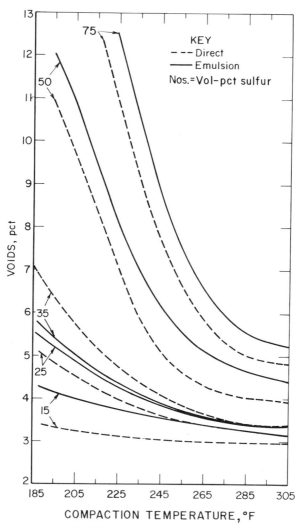

Figure 7. Voids content vs. compaction tempera-
ture—volcanic aggregate

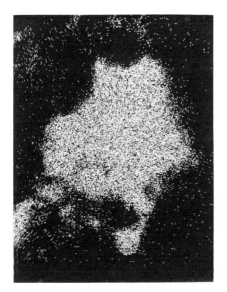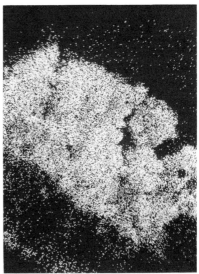

*Figure 8.   Sulfur distribution in asphalt paving specimens, scanning electron
microscopy (300×). Compaction temperature, 305°F.*

same temperatures as those for conventional paving. No differences in
the workability of materials prepared by the two methods were evident.

The distribution of sulfur in the binders was investigated using
scanning electron microscopy. Samples of paving materials prepared by
both methods at compaction temperatures of 185° and 305°F were
prepared and examined. Sulfur was uniformly distributed throughout the
binder, whether prepared by the emulsified or by the direct-addition
method. Figure 8 shows the sulfur distribution (white) in asphalt paving
specimens from paving made with a 25-vol % (42-wt %) sulfur–asphalt
binder. The left photomicrograph is of direct-substituted and the right
is of emulsified sulfur–asphalt paving. There was no large concentration
of sulfur in the boundary between the aggregate particles and the binder,
indicating that with either preparation method, the sulfur was either
initially soluble or dispersed in the binder and remained dispersed in the
binder until re-precipitation of the sulfur started.

Fatigue tests were made on laboratory-compacted bars of paving
prepared from both volcanic and limestone aggregate with asphalt and
direct-substituted sulfur–asphalt binders. These measurements were made
by the Texas Transportation Institute using five constant levels of stress
amplitude, third point, flexural fatigue loading modes as described in
ASTM STP 561 (*14*). Results from these tests are plotted in Figures 9a
and 9b. The results indicate that the fatigue life of these pavements
under constant stress conditions is a function of the sulfur content. Con-

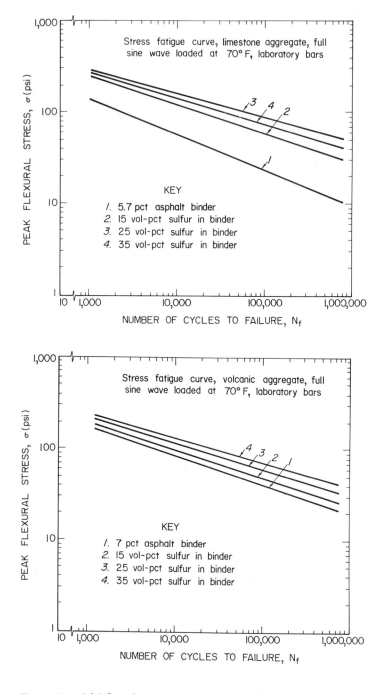

Figure 9. (a) (above) Stress fatigue curve—limestone aggregate.
(b) (below) Stress fatigue curve—volcanic aggregate.

ventional paving materials made using both aggregates exhibited a fatigue life of 1 million cycles (equivalent to a 15-yr life for an average highway) at 10–20-psi constant stress level. In contrast, the materials made with both aggregates and with sulfur–asphalt binders had a similar fatigue life at a 30–40-psi stress level. The improved performance characteristics of sulfur-substituted materials offer promise of more durable pavements and the possibility of thinner road sections, which would conserve both asphalt and mineral aggregate.

Evaluation of sulfur–asphalt binders for paving materials showed no differences in materials prepared using either the emulsified or the direct substitution method in preparing the binder. Therefore, the simpler direct substitution method was used in all subsequent testing.

Upgrading of Marginal Aggregate. LIMESTONE CHAT. The aggregate was a minus ½-in. limestone chat waste material from a limestone quarry. Size gradation of the material shown in Figure 1 does not meet the normal specification for a well graded aggregate. The optimum asphalt binder content was established at 3.7 wt %. Properties of paving material prepared from the aggregate with asphalt and with sulfur–asphalt binders are given in Table III.

Table III. Properties of Limestone Chat Paving Materials

Binder Sulfur (vol %)	Specific Gravity	Voids (%)	Marshall		Dynamic Stiffness (psi × 10^6)
			Stability (lb)	Flow (0.01 in.)	
0	2.479	2.5	2,475	11	0.890
15	2.502	2.2	2,400	12	1.920
25	2.492	2.9	3,345	10	2.210
35	2.489	3.2	7,690	6	2.325
50	2.494	3.6	12,110	6	ND[a]
75	2.511	5.4	12,665	8	ND[a]

[a] ND = not determined.

Temperature sensitivity tests were made on all the products listed in Table III. The results are plotted in Figures 10 and 11. These data are comparable with those obtained with well graded aggregates. Stability and void properties are acceptable for material containing up to 25-vol % sulfur in asphalt binders. Use of a 35-vol % binder is marginal; the stability is high but acceptable. Void contents are excessive when the material is compacted at temperatures below 220°F. Material containing above 35-vol % sulfur appeared to lack flexibility, which is to be expected at these low asphalt binder volume levels. Paving materials prepared from this marginal aggregate and sulfur–asphalt binders had improved stability and dynamic stiffness values.

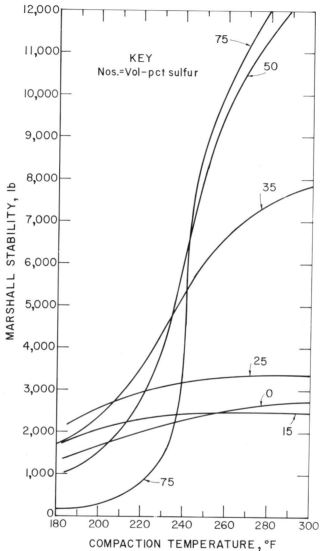

*Figure 10.   Marshall stability vs. compaction tempera-
ture—limestone chat*

DESERT SAND. A maximum Marshall stability of 535 lb was obtained
from paving materials consisting of asphalt and desert sand. By blending
the sand with 7 wt % minus 200-mesh silica flour, an optimum Marshall
stability of 1110 lb was obtained using the gradation shown in Figure 2
with a 6 wt % asphalt binder content. Properties of this material and of
materials made using five levels of sulfur substitution in the binder are
given in Table IV.

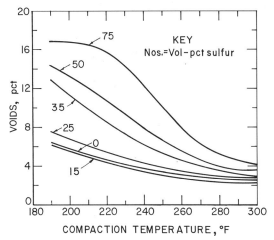

Figure 11.   *Voids content vs. compaction temperature—limestone chat*

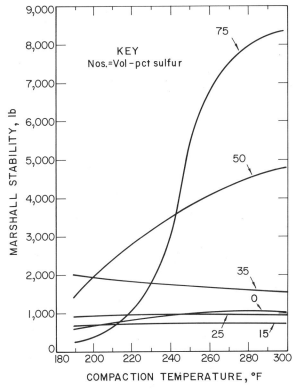

Figure 12.   *Marshall stability vs. compaction temperature—desert sand*

Table IV.   Properties of Desert Sand Paving Materials

Binder Sulfur (vol %)	Specific Gravity	Voids (%)	Marshall Stability (lb)	Marshall Flow (0.01 in)	Dynamic Stiffness (psi × 10⁶)
0	1.965	19.2	1,110	10	0.355
15	1.988	19.0	735	9	0.565
25	2.005	18.7	970	9	0.670
35	2.014	18.5	1,620	11	0.905
50	2.051	18.1	5,600	8	ND[a]
75	2.083	18.1	8,345	8	ND[a]

[a] ND = not determined.

Temperature sensitivity tests were made on these materials, and the results are shown in Figures 12 and 13. From the results it is evident that substitution of 35-vol % sulfur in the binder improves the stability of the sand paving with the same level of voids and flow. The material is workable down to a 190°F compaction temperature. Larger amounts

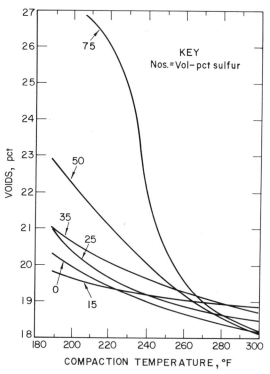

Figure 13.   Voids content vs. compaction temperature—desert sand

Table V. Properties of Volcanic Sand Paving Materials

| Binder Sulfur (vol %) | Specific Gravity | Voids (%) | Marshall | | Dynamic Stiffness (psi × 10⁶) |
			Stability (lb)	Flow (0.01 in.)	
0	1.975	13.7	790	10	0.290
15	1.969	14.6	680	9	0.420
25	1.991	14.2	795	9	0.660
35	1.991	14.7	1,525	9	1.150
50	2.009	14.7	3,990	8	ND^a
75	2.023	15.3	5,750	8	ND^a

^{a} ND = not determined.

of sulfur substitution give higher stabilities but would have to be worked at compaction temperatures above 260°F.

VOLCANIC SAND. The gradation shown in Figure 2 for the volcanic mortar sands was achieved by blending 50% commercial mortar sand with 50% reject slimes washed from the sand. The optimum asphalt binder content for the blend was established at 6 wt %, and this binder level was used in subsequent studies with sulfur-substituted asphalt binders. Properties of paving materials formulated using volcanic mortar sands with both asphalt and sulfur–asphalt binders are given in Table V. Products prepared using the 35-vol % sulfur in asphalt binders were workable at temperatures down to 190°F and had good Marshall stability and flow values.

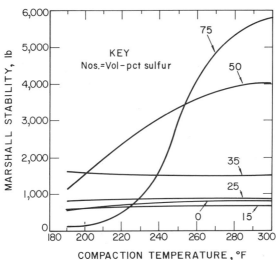

Figure 14. Marshall stability vs. compaction temperature—volcanic sand

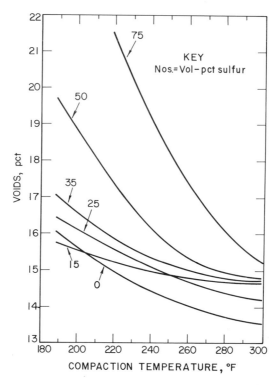

Figure 15. Voids content vs. compaction temperature—volcanic sand

Temperature sensitivity properties of the materials are shown in Figures 14 and 15. High stabilities were achieved at the higher sulfur substitution levels, but compaction was required at temperatures above 260°F.

LIMESTONE SAND. Crushed limestone sand that passed a 1/8-in. screen was used as received. Size gradation of the material is shown in

Table VI.   Properties of Limestone Sand Paving Materials

| Binder Sulfur (vol %) | Specific Gravity | Voids (%) | Marshall | | Dynamic Stiffness (psi × 10⁶) |
			Stability (lb)	Flow (0.01 in.)	
0	2.383	4.6	3,250	10	0.680
15	2.397	4.7	2,835	7	1.105
25	2.332	7.8	3,630	10	1.275
35	2.350	7.4	5,495	8	1.595
50	2.320	9.2	10,820	8	ND^a
75	2.324	10.0	14,000	8	ND^a

^{a} ND = not determined.

Figure 2. The optimum asphalt binder concentration was 4.5 wt %, and this volume of binder was used in all tests. Properties of the paving materials prepared from the limestone with asphalt and with sulfur–asphalt binders are shown in Table VI.

These properties are similar to those obtained with the other sands. Increasing sulfur content in the binder above the 15-vol % level gave increased stability values with flow values remaining constant. Tempera-

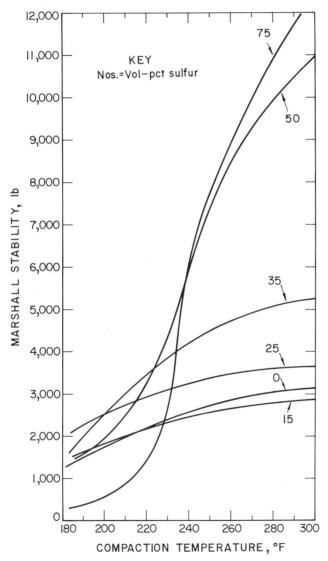

Figure 16.   Marshall stability vs. compaction temperature—limestone sand

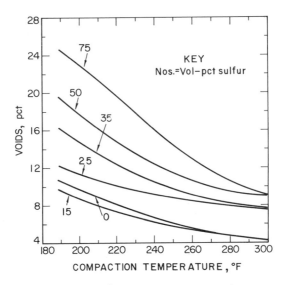

Figure 17.   Voids content vs. compaction temperature—limestone sand

ture sensitivity results are plotted in Figures 16 and 17. Again, good
stability and workability were found in all mixtures up to 35-vol % sulfur
substitution. Higher stabilities were achieved with greater sulfur substitution levels but would require compaction at temperatures above 260°F.

Resistance to the Solvent Action of Fuels. Sulfur–asphalt concretes
were less affected by the solvent action of fuels than was normal asphaltic
concrete. The solvent action of three fuels—gasoline, JP-4 jet fuel, and
No. 2 diesel fuel—was tested on samples of normal asphaltic concrete
and on 15-, 25-, and 35-vol% sulfur–asphalt concretes. Two test methods
developed by McBee (15) were used to determine loss in weight and
loss of stability on drip and total immersion testing of Marshall compaction
samples.

Gasoline was the most active solvent, causing a 19% weight loss
of asphaltic concrete compared with a 4% loss for 35-vol % sulfur–
asphalt concrete using immersion testing. Jet and diesel fuels each had
solvent actions that resulted in a 6% weight loss with normal paving
compared with a 1% loss with 35-vol % sulfur–asphalt concrete.

Marshall stability of asphalt concrete dropped 72% after immersion
testing in gasoline compared with only a 21% loss with 35-vol % sulfur–
asphalt concrete. Jet and diesel fuels had a lesser effect on the Marshall
stabilities than did gasoline. The solvent effect on sulfur–asphalt concrete
materials decreased with increasing sulfur content in the asphaltic binder
in the 0–35-vol % substitution range. The greater resistance of sulfur–

asphaltic concretes to corrosion by fuels makes them attractive for use in areas where fuel spillage is a problem.

Summary of Laboratory Investigations. The laboratory evaluation tests showed that sulfur could be substituted for part of the asphalt binder in the preparation of paving materials both with standard graded aggregates and with marginal aggregates. No differences were observed in the properties of paving materials made with direct-substituted or emulsified sulfur–asphalt binders using type IVb limestone or volcanic aggregates at equal sulfur substitution levels. With these aggregates, up to 52 wt % of the asphalt could be replaced with sulfur with no detrimental effects on the properties such as Marshall stability and flow, compaction temperature limits, fatigue, or dynamic stiffness. There was no problem with segregation of sulfur in the materials.

Successful upgrading of marginal aggregates was achieved by combining selective size gradation of sands with the use of sulfur–asphalt binders. Desert sands and mortar sands were upgraded to Marshall stabilities suitable for heavy-duty pavements using a 35-vol % sulfur–asphalt binder. Offgrade limestone chat and limestone sands, when mixed with sulfur–asphalt binders, produced paving materials equivalent to products prepared with normal aggregates and asphalt binders.

Larger Scale Evaluation Program

Test Strips. Larger batches of direct-substituted sulfur–asphalt paving mixtures were prepared with type IVb aggregate materials. Mixtures containing 0-, 15-, 25-, and 35-vol % sulfur in the asphalt binder were

Figure 18.   Mixers for preparation of sulfur–asphalt paving materials

prepared in 650-lb batches in either a heated Essick mortar or a Barber-Greene model 804 Mixall at 280°–300°F. Both mixers are shown in Figure 18. After mixing and sampling, the material was used to pave an 18-sq ft section of laboratory roadway to a depth of 3 in. Paving was compacted at 200°–250°F with four passes of a 5-ton vibrating steel roller. The properties of the materials, including void levels of core samples cut from the test sections, are shown in Table VII. The data correlate closely with previous laboratory data obtained on small batches of material. Slightly higher stability values were obtained on materials mixed in the mortar mixer.

The values shown in the table indicate that the material compacts to the desired level with a 5-ton vibrating compactor. Surface skid resistance of the sections was measured using a California portable skid tester in accordance with California test method 342 D. The skid values are comparable with published values for fresh-open or dense-graded asphaltic concretes. The friction values increased with increasing sulfur concentration.

Test strips, 8 sq ft in area and 3 in. deep, were prepared from 300-lb batches of paving using the four types of marginal aggregate materials with the size gradations shown in Figures 1 and 2. Four paving strips of each material were prepared using 0-, 15-, 25-, and 35-vol % sulfur-substituted asphalt binders. The binder content for each type of aggregate material was the same as that used in the laboratory evaluation

Table VII. Properties of Sulfur Asphalt Paving Test Strips

Binder Sulfur (vol %)[a]	Marshall		Voids (%)		Coefficient of Friction
	Stability (lb)	Flow (0.01 in.)	Lab.	Core	
0 (L)[b]	2,080	8	2.6	4.5	0.34
0 (V)[b]	2,235	10	2.5	4.5	0.29
15 (L)[b]	2,150	8	2.8	2.4	0.32
15 (L)[b]	2,455	8	3.4	3.3	0.33
15 (V)[b]	3,165	13	2.4	2.9	0.34
15 (V)[c]	4,360	12	2.9	4.3	0.33
25 (L)[b]	2,580	8	2.5	3.6	0.44
25 (L)[c]	3,220	10	2.6	2.3	0.39
25 (V)[b]	4,125	11	2.9	3.2	0.35
25 (V)[c]	4,895	9	2.8	3.8	0.33
35 (L)[b]	4,560	8	2.6	3.4	0.39
35 (V)[b]	5,585	9	3.3	3.5	0.37
35 (V)[c]	7,015	8	3.2	4.9	0.43

[a] (L) = type IVb limestone aggregate, (V) = type IVb volcanic aggregate.
[b] Prepared in Barber–Greene Mixall.
[c] Prepared in Essick mortar mixer.

Table VIII.   Properties of Sulfur Asphalt Paving Test Strips Using Marginal Aggregates

Binder Sulfur (vol %)[a]	Marshall		Voids (%)		Coefficient of Friction
	Stability (lb)	Flow (0.01 in.)	Lab.	Core	
0 (LC)	2,150	8	2.0	8.1	29
15 (LC)	2,005	8	2.0	5.7	34
25 (LC)	3,055	10	2.2	6.2	34
35 (LC)	7,020	4	2.8	8.9	32
0 (DS)	1,490	10	17.0	22.9	22
15 (DS)	1,040	9	18.0	22.7	20
25 (DS)	835	7	16.6	21.8	24
35 (DS)	1,430	7	18.7	23.4	35
0 (VS)	960	8	13.3	15.5	20
15 (VS)	530	9	13.0	17.8	30
25 (VS)	775	8	14.7	19.7	30
35 (VS)	1,360	7	12.8	16.1	29
0 (LS)	3,900	9	6.7	12.6	30
15 (LS)	3,240	6	6.7	20.8	38
25 (LS)	7,550	3	6.8	21.1	40
35 (LS)	9,125	6	8.6	21.5	48

[a] LC = limestone chat; DS = desert sand; VS = volcanic sand; LS = limestone sand.

tests. The paving materials were mixed in the heated Essick mortar mixer at 280°–300°F, sampled, and compacted at 230°–270°F with a vibratory compaction shoe. The properties of the test strips are given in Table VIII.

The properties of the paving material test strips were similar to those of the laboratory test program. One general difference was that all the larger batches of the marginal aggregates with sulfur-substituted binders produced mixtures that were drier than the laboratory evaluation samples and were not as amenable to placement and compaction. This indicates that a slightly larger binder volume may be necessary for field mixing and placing of these materials. This was indicated also by the lower flow values obtained on the field test materials with 35-vol % sulfur-substituted paving.

Road Testing. The direct-substituted sulfur–asphalt paving material was tested under heavy traffic conditions in cooperation with the city of Boulder City, NV. A 100-sq ft replacement patch was made in the center of an intersection using type IVb limestone aggregate, AR 2000 asphalt, and sulfur. Paving materials prepared with 15-, 25-, and 35-vol % sulfur–asphalt binders in the Barber-Greene Mixall were laid in three 4 × 8-ft segments to a depth of 4 in. Paving was laid in a 2-in. base and a 2-in. surface course. The base courses were compacted with a vibrating

Figure 19.   Laying patch of direct-substituted asphalt paving materials

Table IX.   Boulder City Street Patch Data

| Composition (wt %) | | | Amount Used (lb) | Specific Gravity | Voids (%) | Marshall | | Coefficient of Friction |
Sulfur	Asphalt	Aggregate				Stability (lb)	Flow (0.01 in.)	
1.85	5.15	93.00	1600	2.4064	2.7	1565	9	27
2.89	4.52	92.59	1900	2.4096	3.1	2490	11	26
4.14	3.87	91.99	1600	2.4302	2.9	3840	8	31

Figure 20.   Placement and compaction of direct-substituted sulfur–asphalt paving in Ohio

shoe compactor, and the surface courses with a 5-ton double steel roller. Compaction of the surface course with the roller is shown in Figure 19. Properties of the patch material are given in Table IX. The patch has been in service for over a year with no signs of degradation or failure. The properties of the three levels of sulfur–asphalt patch materials compare well with previously determined data for these mix designs.

A 500-ft road section of direct-substituted sulfur–asphalt pavement was placed by Standard Oil Co. of Ohio in Ohio during August 1975. The section consisted of a 20-ft wide, 3-in. deep overlay paving using a limestone rock and sand aggregate with a 52 wt % sulfur–asphalt binder. The materials were mixed in a 4000-lb pugmill and placed and compacted with conventional paving equipment. Placement and rolling of the material are illustrated in Figure 20. The performance of this test section is being monitored and at present shows no signs of wear or degradation.

Conclusions

The direct addition of sulfur and asphalt in the mixer as binder materials was shown to be practical in the preparation of dense-graded asphaltic concrete in both laboratory and field testing. A laboratory evaluation of sulfur–asphalt concretes prepared by both the emulsion and the direct addition methods showed no differences in the properties of the paving materials. The substitution of up to 35 vol % (52 wt %) of the asphalt with sulfur gave materials with properties equal or superior to those of normal asphaltic concretes, and these materials could be prepared in mix plants, laid with pavers, and compacted with rollers that are commonly used in the paving industry. Higher stability materials that might be suitable for base courses can be produced by replacing the asphalt with amounts greater than 52 wt % of sulfur, and compacting at temperatures over 240°F.

By using the increased stability realized with a combination of size grading of sands and a 35-vol % sulfur–asphalt binder, marginal sand aggregates were upgraded to prepare sulfur–sand–asphalt paving that would meet the requirements for heavy-duty highways. The use of larger amounts of sulfur in the binder with marginal aggregates makes the resulting concretes promising for base courses if they can be placed and compacted at temperatures above 250°F.

Sulfur replacement of up to 52 wt % of the asphalt in binders has been possible, with no detrimental effects noted to date. Also, sulfur–asphalt concretes are more resistant to the corrosive action of fuels than are normal asphaltic concretes.

Literature Cited

1. U.S. Department of the Interior. First Annual Report of the Secretary of Interior under The Mining and Mineral Policy Act of 1970, **PL. 91-631**, pp. 34, 104, U.S. Government Printing Office, Washington, D.C., March 1972.
2. U.S. Department of Transportation, Federal Highway Administration, Press Release FHWA **28-75**, April 3, 1975.
3. Burgess, R. A., Deme, I., "Asphalt Paving Mixes," ADV. CHEM. SER. (1975) **140**, 85.
4. Sullivan, T. A., McBee, W. C., Rasmussen, K. L., "Studies of Sand–Sulfur–Asphalt Paving Materials," *U.S. Bur. Mines Rep. Invest.* (1975) **8087**.
5. Saylak, D., Galloway, B. M., "The Use of Sulfur in Sulfur Asphalt Aggregate Mixes," *Interam. Conf. Mater. Technol.*, 4th (July 1975) 636–646.
6. Bencowitz, I., Boe, E. S., "Effect of Sulfur Upon Some of the Properties of Asphalts," *ASTM Proc.* (1938) **38**, Part II, 539–550.
7. Garrigues, C., Vincent, P., "Sulfur/Asphalt Binders for Road Construction," ADV. CHEM. SER. (1975) **140**, 130.
8. Kennepohl, G. J. A., Logan, A., Bean, D. C., "Sulfur–Asphalt Binders in Paving Mixes," Proceedings of the Canadian Sulfur Symposium, Calgory, Alberta, May 1974, pp Q1–16.
9. "Sulfur Being Tested as Road Asphalt Substitute," The Free Press, Diboll, Texas (September 11, 1975) **22** (37) 1.
10. McBee, W. C., Sullivan, T. A., "Utilization of Secondary Sulfur in Construction Materials," *Proc. Miner. Waste Util. Symp.*, 5th (1976) 39–52.
11. *Book of ASTM Stand.* (1976) Part 15.
12. The Asphalt Institute, "Construction Specifications for Asphaltic Concrete and Other Plant-Mix Types," Specifications Series No. 1 (55-1) 4th ed., November 1969.
13. Schmidt, R. J., "A Practical Method for Determining the Resilient Modulus of Asphalt Treated Mixes," Highway Research Record **404**, National Academy of Sciences, Washington, D. C., 1972.
14. Irvin, L. H., Galloway, B. M., "Influence of Laboratory Test Method on Fatigue Test Results for Asphalt Concretes," Fatigue and Dynamic Testing of Bituminous Mixtures, ASTM STP **561**, American Society for Testing Materials, pp 12–46, 1974.
15. McBee, W. C., Sullivan T. A., "Improved Resistance of Sulfur Asphalt Paving to Attack by Fuels," *IEC Prod. Res. Dev.* (March 1977) **22** (1).

RECEIVED April 22, 1977. References to specific companies, brands, and trade marks are made for identification only and do not imply endorsement by the Bureau of Mines.

9

Construction and Performance of a Sulfur–Asphalt Road in Texas

S. MORGAN PRINCE

State Department of Highways and Public Transportation,
District 11, Lufkin, TX 75901

After many years of laboratory study, sulfur–asphalt binders, in which sulfur substituted for 30% of the asphalt, were used to construct a highway in the U.S. Variables included in the designed experiment were thickness, aggregate type, and the binder content of the mix. Except for the manufacture of the binder, the highway was constructed using conventional equipment. Although sulfur is nontoxic, the environmental impact of sulfur paving material was studied. Emissions of H_2S and SO_3 were extremely low, usually less than 1 ppm. Analysis of soil and water samples indicated that leaching of sulfur from the highway has not occurred. Performance data do not show yet that statistically significant differences exist between the sulfur–asphalt binder and the asphalt binder used as a control.

Although the idea of using sulfur in asphalt is older (*1*), the first practical tests apparently were carried out in Texas in the early 1930s. Bacon and Bencowitz (*2, 3, 4*) made a careful study of the treatment of 17 different asphalts with sulfur at 150°C. They reported the effect of sulfur on asphalt binder and on the mixes. In 1934, a private road was constructed in which sand–limestone mixes were used as aggregates. The work was relatively successful, and mixes prepared from sand and sulfur–asphalt binders performed better than those made with sand and asphalt alone. However, because asphalt was readily available and priced lower than sulfur, the developmental efforts were discontinued for nearly three decades. For example, in a 1971 review of the subject, Fike (*5*) reported that only two papers (*6, 7*) describing the beneficial aspects of adding sulfur to asphalt were published during 1950–70.

0-8412-0391-1/78/33-165-161$05.00/0

During the early 1970s the potential advantages of sulfur–asphalt over conventional binders were recognized, and research was initiated by many organizations (8, 9, 10, 11, 12). In contrast to much of the early work, research during the 1970s was designed to use the physical properties of sulfur to enhance the existing excellent properties of asphalt. The investigators claimed that binders containing both sulfur and asphalt might permit sand and other low-quality aggregates to wholly or partially replace conventional aggregates, such as crushed stone. Investigators also demonstrated that the amount of asphalt required in the paving mix can be significantly reduced where sulfur is used.

Highway District 11 is located in East Texas, and the area does not have an abundant supply of conventional aggregates. Of the materials available locally, sand is the most plentiful. These local sands were first combined with asphalt in a hot-mix plant in 1962 and were evaluated on U.S. and State highways. From this experience, it was concluded that hot-hand asphalts merited consideration as base materials. Quite often two sands (usually 100% passing the 40 sieve) are combined to obtain the proper gradation. When it was subsequently learned that the Texas Transportation Institute (13) was evaluating the sulfur–asphalt binders, efforts were initiated to field test these materials.

This report will describe the design, construction, and performance of a test highway in which a sulfur–asphalt binder was used with sand alone and sand–gravel aggregates. The program was quite comprehensive and included the study of other variables such as thickness and amount of binder.

Table I. Physical Properties of Sulfur–Asphalt Binders

Blend of				
asphalt[a]	100	75	70	60
elemental sulfur		25	30	40
Specific				
gravity (16°C)	1.0172	1.16	1.19	1.27
Softening				
point (°C)	46	51	48	53
Penetration				
25°C	83	81	76	76
100g, 5sec				
4°C	17	20		12
Viscosity (poise)				
25°C	960,000	1,300,000	1,840,000	3,100,000
60°C	1,280	1,200	962	1,330
120°C	4.98	3.01	2.92	3.40
135°C	2.64	1.39	1.55	1.50
150°C	1.73	0.84	0.80	1.21

[a] Texas AC-20, Port Neches, TX.

Table II.   Dry Sieve Analyses of Aggregate Used

Sieve Size	Sand % Passing by Weight	Gravel % Passing by Weight	65% Gravel 35% Sand % Passing by Weight
3/8″		99	99
No. 4		44	64
No. 10		3	37
No. 20	100		35
No. 40	88		31
No. 80	37		13
No. 200	11		4

Laboratory Design Data

Sulfur–asphalt binders can be prepared by various mechanical means. One conventional method is to combine liquid sulfur and asphalt at 285°–300°F in a Gifford Wood colloid mill. A rotor stator gap setting of 0.02 in. at 7000 rpm for 8 min will prepare satisfactory emulsions. This emulsion is immediately mixed with preheated aggregate. The laboratory binder was prepared by TTI (Texas Transportation Institute) scientists in cooperation with SNPA (Société Nationale des Pétroles d'Aquitaine) scientists and is believed to be comparable with the binders prepared by the turbine in the field trials. The need for dispersing sulfur in the asphalt is discussed by Garrigues (9) and by Kennepohl et al., Deme, and McBee et al. elsewhere in this volume.

The physical properties of several sulfur–asphalt blends were determined in the laboratory and are illustrated in Table I. The softening points and penetration of the asphalt were relatively unchanged. The viscosity of the sulfur–asphalt blend is generally lower at all temperatures and is significantly less at 135°C. The aggregates used in the trial are described in Table II.

The properties of the mix were determined for sulfur–asphalt binder contents at 4–8% by weight and were compared with the mix using asphalt binder. The results were reported by TTI (14) and are summarized in Table III. A more comprehensive review of the properties of sulfur–asphalt binders was presented by Garrigues et al. (9). In general, the sulfur–asphalt mixtures gave higher Marshall stabilities than conventional asphalt mixes when used with either sand–gravel aggregates or with sand alone.

Construction

During the spring of 1975, the SDHPT (State Department of Highways and Public Transportation) held a meeting at the District 11 office.

Table III. Comparison of Marshall and Hveem Stability Results for
Mixture Designs of Various Type Binders and Aggregates

Type of Binder	Type of Aggregate	Binder Content (% by wt)	Marshall (lb)	Hveem (%)	Air Voids (%)
Sulfur– asphalt[a]	sand	5.0	1280	30	17
		5.5	2080	33	16
		6.0	2670	31	13
		7.0	1650	29	12
		8.0	1790	31	10
	sand–gravel	5.0	2250	44	8
		5.4	2600		6
		5.5		43	
		6.0	2250	37	4
		7.0	2270	37	2
Asphalt[b]	sand	5.4	1460	23	14
		5.5	1540	23	16
	sand–gravel	4.5	1680	35	8
		5.0	1740	32	6

[a] 30% Sulfur by weight.
[b] Texaco AC-20, Port Neches, TX.

This meeting was attended by representatives of various groups who agreed to cooperate in the construction of test sections on US 69 northwest of Lufkin. Two new lanes were being constructed to convert the two-lane facility into a four-lane highway. Traffic volume averages 6000 vehicles per day, 15% of these are 18-KIP axle-load trucks. SNPA per-

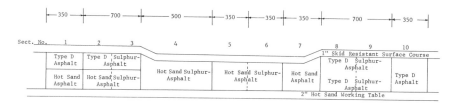

Figure 1. Layout of SNPA sulfur–asphalt binder pavement test, US highway 69, Lufkin, TX

Figure 2. SNPA's mobile sulfur and asphalt blending equipment

mitted the contractor to use its equipment to blend sulfur and asphalt. This equipment was shipped from France to the U.S. in June 1975. The Sulphur Institute and the Federal Highway Administration also contributed funds to assist the Texas SPHPT to construct and evaluate the performance of the highway. Contributions of others are noted at the end of this chapter.

The design of the test sections is illustrated in Figure 1. Binders were selected which contained 30% sulfur by weight. As high a substitution ratio as possible was desired to minimize asphalt usage, and, at the time,

Figure 3. Conventional hot-mix paver and haul equipment used to place both the sulfur–asphalt and non-sulfur-containing mixes

Table IV. Comparison of Design Mixtures and Field
Laboratory Hveem Stability Test Results

Binder	Aggregate	Binder Content (% by wt.)	Hveem Stability (%) Design	Field
Sulfur–	sand	5.5	33	
asphalt[a]		6.0	31	34
		6.2		32
		7.0	29	29
		7.5		30
	sand–gravel	4.8		40
		5.0	44	
		5.3		39
		5.5	43	
		5.65		44
		6.0	37	37
Asphalt[b]	sand	5.4	23	22
		5.5	23	22
	sand–gravel	4.5	35	
		5.0	32	32

[a] 30% Sulfur.
[b] Texaco, AC-20, Port Neches, TX.

little data were available on substitutions above the 30% level. The colloid mill operated by SNPA personnel is shown in Figure 2. After the sulfur–asphalt binder mixes were prepared at the plant, trucks and pavers were operated in a conventional manner. These operations took place in September 1975 (just five months after the first meeting concerning the test section) and have been presented in detail by Izatt (15). Figure 3 illustrates the laying of the material at the site.

Table V. Field Performance for Preliminary (October 1975),

Binder Type	Aggregate Type	Binder Content (% by wt)	Air Voids Prel.	Sec.	Tert.	Hveem Stability (%) Prel.	Sec.	Tert.
Sulfur–	sand	6.0	23	21	20	21	24	22
asphalt[a]		6.35	21	20	22	20	23	20
		7.1	20	20	16	24	22	23
	sand–gravel	4.8	8	11	7	22	27	25
		5.65	11	8	4	18	31	32
Asphalt[b]	sand	5.4	21	22	22	15	21	19
	sand–gravel	4.8	8	9	5	21	28	31

[a] 30% Sulfur.

No difficulties were encountered at the plant or at the paving site. The procedures used for preparing and laying the sulfur–asphalt mixes were identical to those used for the non-sulfur-containing mixes. Hveem stabilities of the plant-mixed materials as measured in the field laboratory compared quite well with those of the design mixes as shown in Table IV.

Performance

The test sections were not opened to regular traffic until June 1976; therefore they have been under traffic for only 10 months as of this report. A preliminary series of samples was taken one month after the test sections were completed (October 1975), a second series soon after the sections were opened to traffic (August 1976), and a third series of samples in April 1977. Additional test series are planned at six-month intervals until at least six series of data are obtained. The results of the preliminary and secondary testing phases have been reported by Texas Transportation Institute (*16*) and are summarized in Table V.

The data indicate that the Hveem stabilities of the mixes have increased between preliminary testing (October 1975) and the secondary testing (August 1976). The Marshall stabilities displayed a significant increase while accompanied by only slight change in Marshall flow values. These increases are presumably caused by compaction from the traffic, and further increases are likely to occur. None of these values are as high as those obtained for the same materials during the design stages or the field laboratory test, suggesting that compaction during construction was less than anticipated. In general, the sulfur–asphalt mixes and the asphalt mixes are presently displaying comparable values.

Secondary (August 1976), and Tertiary (April 1977) Testing Phases

Marshall Stability (lb)			Marshall Flow (0.01 in.)			Resilient Modulus (x 10^6 psi)		
Prel.	Sec.	Tert.	Prel.	Sec.	Tert.	Prel.	Sec.	Tert.
169	340	960	13	12	12	0.13	0.28	0.24
152	613	730	15	14	13	0.14	0.36	0.18
137	512	850	18	13	13	0.20	0.37	0.26
425	485	1100	15	15	13	0.29	1.22	0.52
200	720	1190	14	12	14	0.26	0.45	0.60
352	645	620	14	14	15	0.16	0.31	0.24
388	554	1010	16	14	12	0.24	0.84	0.59

[b] Texaco AC-20, Port Neches, TX.

Benkelman beam rebound deflection measurements have been obtained during each of these testing phases. The SDHPT Dynaflect Unit has been used to obtain dynamic stiffness measurements. These results also indicate that the sulfur–asphalt mixes and the asphalt mixes are displaying comparable characteristics.

Visually, all test sections appear to be performing satisfactorily. We evaluated the surface riding characteristics by use of the Mays Ride Meter which indicated that the various sections possess essentially the same degree of roughness.

Environmental and Safety Aspects

Although sulfur is considered nontoxic to humans (17, 18), it can cause eye irritation. In addition, sulfur could react with asphalt to form hydrogen sulfide, sulfur dioxide, or other gaseous emissions. Sulfur is also an important plant nutrient. However, as with other nutrients, an excess can cause undesirable effects on vegetation. The SDHPT worked with the U.S. Bureau of Mines, TTI, The Sulphur Institute, and Stephen F. Austin State University to study the environmental impact of sulfur.

Hydrogen sulfide emissions were monitored at the sulfur–asphalt mixing equipment at the hot-mix plant and at the paving site. A Houston-Atlas ambient H_2S detection system, furnished and operated by the U.S. Bureau of Mines, was used as well as portable equipment such as a Rotorod Gas Sampler and Drager tubes. The emissions at some of the critical areas of the hot-mix plant and paving site are given in Table VI.

Based on the recommended maximum allowable concentration (MAC) values (Table VII), the H_2S concentrations measured in the areas most frequented by construction personnel are well below the critical threshold. The highest H_2S concentrations were measured inside the sulfur–

Table VI. Emissions at Some Critical Areas of Hot-Mix Plant and Paving Site[a]

Location	H_2S Concentration (ppm)
Batch weight control room	< 0.5
Vicinity of the colloid mill	negligible
Inside opening of S/A emulsion storage tank	15
Pugmill platform	< 1.0
Over dump truck body	< 1.0
Paver hopper during truck dumping	< 1.0
Paver hopper during paving	negligible

[a] Based on the recommended maximum allowable concentration.

Table VII. Hydrogen Sulfide Concentration Levels

H_2S Concentration (ppm)	Environmental Impact
.02	odor threshold value
5–10	suggested MAC[a]
20	MAC
70–150	slight symptoms after exposure of several hours
170–300	maximum concentration that can be inhaled for 1 hr without serious consequences
400–700	dangerous after continuous exposure of 30 min–1 hr
600	fatal with exposure greater than 30 min

[a] Maximum allowable concentration

asphalt emulsion surge tank, and this, as well as all tanks holding chemicals, is usually sealed off from all but highly trained personnel.

During the paving operation, the paver operator experienced eye irritation resulting from colloidal sulfur. This was particularly objectionable when there was no breeze. Therefore paver operators should wear goggles to eliminate this discomfort.

Sulfur dioxide emissions were generally the same as H_2S. Only the holding tank registered concentrations exceeding 1 ppm. Ref. *15* describes these findings.

Leaching into the Soil

Sulfur is required for plant growth and ranks in importance with nitrogen and phosphorus in the formation of protein. Agricultural uses of sulfur and its compounds have been reviewed by The Sulphur Institute (*19, 20, 21*). Soils in the Lufkin area are deficient in sulfur and, as a result, leaching of sulfur from the pavement is not likely to be harmful, even if it were to occur. However, leaching might be undesirable in some regions. As a result, the SDHPT and The Sulphur Institute are working with scientists at Stephen F. Austin State University to test periodically the soil and water runoff to determine if, or to what extent, leaching will occur. Although the program is just over one year old, thus far there is no evidence to suggest that leaching will be a problem.

Conclusion to Date

The limited data generated to date indicate that the performance of sulfur–asphalt binder mixes is comparable with that of the asphalt mixes and that sulfur can be used to replace part of the asphalt used in paving mixtures. Further data must be collected and analyzed before we can conclude that use of sulfur–asphalt binders in combination with low-quality aggregates can replace conventional aggregates.

Acknowledgments

The author wishes to acknowledge the assistance of Frank D. Gallaway, District Engineer (deceased) and appreciated his confidence that all organizations could accomplish their tasks within the needed time frame. We wish also to thank especially Bob M. Gallaway, Texas Transportation Institute, and H. L. Fike and J. O. Izatt, The Sulphur Institute, for their contributions.

The contributions of the Federal Highway Administration, Moore Brothers Construction Co., Société Nationale des Pétroles d'Aquitaine, Texas Transportation Institute, U.S. Bureau of Mines, Texasgulf, Inc., Robertson Tank Lines, Inc., and the Sulphur Institute were greatly appreciated.

Literature Cited

1. Abraham, H., "Asphalts and Allied Substances," Vol. 1, p. 80; pp. 178–179; Vol. 3, pp. 12, 87; 6th ed., Van Nostrand, New York, 1960.
2. Bacon, R. F., Bencowitz, I., "Method of Paving," U.S. Patent 2,182,837, Dec. 12, 1939.
3. Bacon, R. F., Bencowitz, I., Book ASTM Stand. (1938) 38 (2), 539.
4. Bencowitz, I., Boe, E. S., "Effect of Sulphur Upon Some of the Properties of Asphalts," ASTM Proc. (1938) 38 (2).
5. Fike, H. L., Adv. Chem. Ser. (1972) 110, 208.
6. Metcalf, C. T., "Bituminous Paving Composition," British Patent 1,076,-866, July 26, 1967.
7. Speer, T. L., "Bituminous Pavement," U.S. Patent 3,239,361, March 8, 1966.
8. Deme, I., Hammond, R., McManus, D., "The Use of Sand-Asphalt-Sulphur Mixes for Road Base and Surface Applications," Proc. Can. Tech. Asphalt Assoc. (November 1971) IVI.
9. Garrigues, C., Vincent, P., Adv. Chem. Ser. (1975) 140, 130.
10. Kennepohl, G. J. A., Logan, A., Bean, D. C., "Conventional Paving Mixes with Sulphur-Asphalt Binder," Proc. Annu. Meet. Assos. Asphalt Paving Technol., Phoenix, AZ (February 1975).
11. Saylak, D., Gallaway, B. M., Epps, J. A., "Recycling Old Asphalt Concrete Pavements," Proc. Miner. Waste Util. Symp., 5th (April 1976).
12. Sullivan, T. A., McBee, W. C., Rasmussen, K. L., U.S. Bur. Mines Rep. Invest. (1975) 8087.
13. "The Extension and Replacement of Asphalt Cement with Sulphur," Texas A&M Research Foundation Contract RF 3259.
14. "Sulphur/Asphalt Mixture Design and Construction Details—Lufkin Field Trials," Texas Transportation Institute, Interim Reports 512-1, Study No. 1-10-75-512 (January 1976).
15. Izatt, J. O., "Sulphur-Asphalt Pavement Binder Test, US 69—Lufkin, Texas, A Construction Report," The Sulphur Institute, September 1975.

16. "Post Construction Evaluation of Sulphur-Asphalt Pavement Test Sections," Texas Transportation Institute, Interim Report, Study No. 1-10-75-512.
17. *Manuf. Chem. Assoc. Chem. Saf. Data Sheet* (1959) **SD-74**.
18. National Safety Council. Handling Liquid Sulphur. Data Sheet **592**, Chicago, Ill., 1966.
19. "Sulphur—The Essential Plant Nutrient," Sulphur Institute, Washington, D.C., 1975.
20. Burns, G. R., "Oxidation of Sulphur in Soils," *Sulphur Inst., Tech. Bull.* (1968) **13**.
21. Bixby, D. W., Beaton, J. D., "Sulphur-Containing Fertilizers, Properties and Applications," *Sulphur Inst. Tech. Bull.* (1970) **17**.

RECEIVED April 22, 1977.

10

Sulfur as an Asphalt Diluent and a Mix Filler

IMANTS DEME

Shell Canada Limited, Oakville Research Centre, P.O. Box 2100,
Oakville, Ontario L6J 5C7, Canada

Sulfur is used currently in two types of asphalt paving proc-
esses. One uses a low amount of sulfur as an asphalt diluent
in conventional asphalt concrete mixes. The other process,
Thermopave, uses higher sulfur contents where the excess
sulfur performs as a mix filler, permitting attainment of
high-quality paving materials using inexpensive sands. The
differences between the two processes and the roles of the
sulfur in each type of paving mix are described emphasizing
microscopic studies of mixes which show that sulfur is
readily dispersed by aggregate shear forces during mixing
in the plant pugmill. Pre-emulsification of sulfur–asphalt
binders is not essential in either process. The emulsified
portion of sulfur performs as an asphalt extender while any
excess sulfur performs as a mix filler or stabilizer.

Various researchers have attempted to use sulfur in asphaltic mixes
over the years; however, only recently has it become practical to do so.
This is attributable partly to the development of new technology and
partly to a major shift in the price of road building materials. The prices
of asphalts and aggregates have escalated substantially within the last
decade, whereas the price of sulfur, although cyclic, has followed a down-
ward trend. The decrease in sulfur cost is attributable to the increase in
the available sulfur reserves associated with its removal from natural gas
and the clean-up of industrial fuels and stack gases. The surplus from
the latter source is expected to increase in heavily industrialized areas as
environmental controls become more restrictive in the future (1).

There are two reasons for considering the use of sulfur in asphalt
mixes: to improve mix quality and/or to reduce mix cost. A small
quantity of sulfur may be dispersed in asphalt, permitting a cost reduction of binder for hot-mix made with conventional graded aggregates.
A number of individuals and organizations are currently involved in the

0-8412-0391-1/78/33-165-172$05.00/0

testing and commercialization of processes using sulfur as a diluent for asphalt binders: Société Nationale des Pétroles d'Aquitaine (SNPA) (2), Gulf Canada Limited (3, 4, 5), Pronk-Sulphur Development Institute of Canada (6), U.S. Bureau of Mines, Texas Transportation Institute, and D. Y. Lee (7). Some of the above cite a need for specialized equipment to disperse the sulfur in the asphalt while some cite improvement in certain mix properties such as Marshall stability, depending upon the amounts and relative proportion of sulfur to asphalt used. Most of the conventional construction equipment may be used to process, transport, and place these mixes.

The Thermopave process, developed by Shell Canada Ltd. (8, 9, 10, 11, 12), uses significantly greater quantities of sulfur than the sulfur-diluted process above. When liquid, the sulfur in Thermopave increases the mix workability to a point where it is placed without roller compaction. The liquid sulfur also conforms to the shape of the void spaces in the mix. When it cools, the principal action of the solidified sulfur is to key in the asphalt-coated aggregate particles. This yields high-quality mixes, even with inferior aggregates and sands. The manufacture of high-quality Thermopave mixes with ungraded sands, in lieu of conventional graded aggregate hot-mix, offers a cost saving in many areas where sands are readily available and aggregates are expensive. Sulfur addition to other specified aggregate formulations permits the attainment of mixes with distinctive characteristics, applicable to a variety of specialized uses as described in Refs. 8, 10, and 12. While hot-mix plants may be adapted readily for Thermopave processing by adding a liquid sulfur supply system, specialized equipment is required for mix transport and placement. Various features of the Thermopave process are covered by patents and patent applications in a number of countries.

Despite numerous publications about Thermopave and sulfur-extended binder processes, a considerable amount of confusion exists about the fundamental concepts governing these two complementary systems. This chapter, therefore, demonstrates how they are related and outlines the mix properties which are most significantly influenced by sulfur addition.

Extension of Asphalt with Sulfur

A number of researchers have studied the effects of treating asphalts with sulfur. The solubility of sulfur in asphalt increases with temperature but is quite low at temperatures below 149°C (the maximum safe temperature for practical mix handling operations to avoid hydrogen sulfide formation). Attempts have been made to form homogeneous dispersions of sulfur in asphalt by mixing, emulsification, and pumping action. The general concensus is that sulfur dispersion has a favorable action on asphalt (*13, 14, 15, 2, 7*). Some of the beneficial properties indicated are increase in penetration value, lowering of the Fraas breaking point temperature, lowering of the softening point temperature, improvement of the penetration index, and an increase in binder ductility.

Controversy exists over the nature of the chemical reaction of sulfur with asphalt, the role of the sulfur remaining in colloidal solution/dispersion in the asphalt, and the maximum permissible sulfur concentration in the asphalt which will yield a stable long-term binder. Several investigators, such as Kennepohl et al. (*4*), have shown that on a long-term basis, approximately 20% of the sulfur remains in a dissolved and/or a dispersed state as part of the binder. Figure 1, prepared from their differential scanning calorimeter data, indicates that excess sulfur, i.e., free sulfur above approximately 20 wt %, solidifies to a crystalline state, ceasing to perform as a binder.

High shear during mixing is essential to obtain good sulfur dispersion and a stable binder. Certain agencies specify the use of colloid mills to achieve this. On the contrary, our experience with Thermopave and other sulfur asphalt products has been that sulfur is well dispersed by aggregate shearing action during mixing in the pugmill of a hot-mix plant. Nevertheless, laboratory and field studies were undertaken to assess the influence on sulfur dispersion of several aggregate variables, the

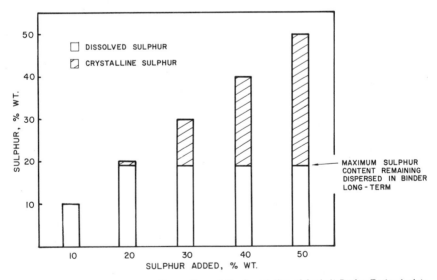

Proceedings of the Association of Asphalt Paving Technologists

Figure 1. Relative amounts of crystalline and dissolved sulfur in sulfur–asphalt binders as determined by differential scanning calorimetry (4)

Table I. Gradation of Materials Used in Sulfur Dispersion Study

Percent Passing

Sieve Size		A Graded Aggregate	B Crushed Stone	C Sand	D Glass Beads
ASTM	mm				
½ in.	13.2	100	100		
⅜ in.	9.5	74	78		
¼ in.	6.7	—	21		
No. 4	4.75	48	0	100	
8	2.36	40		97	
16	1.18	30		92	
30	0.600	21		85	100
50	0.300	9		5	1
100	0.150	3		1	0
200	0.075	0		0	
Surface area (cm²/g)		32.5	15.5	57.2	34.8

mixing mode, and mixing time within the framework of formulations commonly used for sulfur-extended binder mixes and Thermopave.

Sulfur Dispersion by Laboratory Mixer. The influence of the following variables on the dispersion of sulfur in asphalt was assessed in the laboratory. The mixes were prepared using 85–100 pen grade asphalt with all of the mix ingredients preheated to 140°–150°C.

(1) Aggregate (for gradation see Table I):
  (a) Spherical glass beads
  (b) Sand
  (c) Crushed stone
  (d) Dense graded aggregate

(2) Mixing sequence (using a small Hobart mixer, model N-50, 80 rpm, flat beater):
  (a) Simultaneous—all ingredients were combined in a single wet-mix cycle
  (b) Regular—aggregate was pre-mixed with asphalt for 30 sec followed by the addition of sulfur and mixing in a second wet-mix cycle for the various time periods described below
  (c) Reverse—opposite order to (2b), with the sulfur pre-mix followed by asphalt addition and mixing

(3) Mixing time (after last mix ingredient added): 10, 30, 60, and 90 sec

(4) Mix types:
  (a) Sulfur-extended binder mix—binder content, 7 wt % mix; binder composition, 1.75 wt % sulfur and 5.25 wt % asphalt; sulfur/asphalt weight ratio, 0.33
  (b) Thermopave mix—asphalt content, 6 wt % mix; sulfur content, 12 wt % mix; sulfur/asphalt weight ratio, 2.0

Slides for microscopic examination of the binders were prepared as follows. A glass slide and a glass rod were heated to approximately 150°C. Hot mix was pressed against the slide and then removed, leaving some binder on the slide. This was smeared with a single stroke of the glass rod to make a thin binder film of variable thickness. The slide was then examined with a high-powered optical microscope using light transmission and adjusting polarization until the definition of the dispersed sulfur droplets was optimum. The initial examination was performed within a few minutes of mix preparation. But later examination over periods up to 48 hr showed no significant changes in the appearance of the binder.

Additionally a sulfur/asphalt dispersion was prepared in a Waring blender by mixing sulfur and asphalt for 5 min at 20,000 rpm. Examined under the microscope, the supercooled dispersed sulfur globules were approximately 4 μ in diameter. This is typical of sulfur dispersions in asphalt produced by SNPA (2) and Gulf Canada Ltd. (3) using colloid mills and provided a basis for assessing the effectiveness of sulfur dispersion by the shearing action of the aggregate during mix processing.

A study of the binders prepared by aggregate shearing during mixing revealed that good dispersion, with only a few sulfur globules larger than 4 μ, could be attained generally at mixing times as short as 10 sec, regardless of the order in which the constituents were combined. This is believed to result from both the low viscosity of the sulfur (easily sheared) and from the high aggregate shear during mixing. The crushed stone mix (Aggregate B in Table I), probably because of its lower specific surface area, was not as effective in dispersing the sulfur. In this case, examination of the binder showed a few sulfur globules considerably larger than 4 μ up to 30 sec mixing but not after 60 sec mixing. Binders from all of the other aggregate mixes showed good sulfur dispersion after mixing for 30 sec. Longer mixing times did not yield a finer dispersion of sulfur, indicating that the optimum mixing time with the last ingredient added was between 10 and 30 sec.

Attainment of fine sulfur dispersions such as described above are crucial to utilization of sulfur as an asphalt diluent. Provided that they remain stable, the binders will exhibit the sulfur-extended binder properties described previously.

A typical Thermopave mix using sand (Aggregate C in Table I) was prepared using the regular mixing sequence with 6 wt % asphalt and 12 wt % sulfur, a considerably higher amount than used in sulfur-extended binders. Examination of the binder phase on thin film portions of the microscope slides showed numerous finely dispersed sulfur particles. The predominant sulfur particle diameter was 4 μ, similar to the size observed for the sulfur-extended binders prepared in the blender. This also indicated that shear by aggregate during mixing was as effective in Thermo-pave mixes as in the sulfur-extended binder mixes.

The number and size of the large sulfur globules increased with the sulfur–asphalt film thickness on the microscope slide. It is postulated that all or most of the sulfur is well dispersed during mixing but that because of the presence of more sulfur than can be held in suspension by the asphalt, agglomeration of the "excess" sulfur takes place soon after mixing which solidifies on cooling and ceases to perform as an effective part of the binder. The rates of sulfur agglomeration and reaction with asphalt are not known, but their joint effect, combined with that from additional changes in the crystal structure of the free sulfur, is reflected by the rate of mix curing, as discussed later.

Any reacted sulfur and the discrete sulfur particles which remain in suspension in the bitumen long term can be expected to perform as asphalt diluent in Thermopave. But no saving in asphalt content can be realized in optimum-quality Thermopave made with sands, relative to conventional graded aggregate hot-mix. This is because sands have a higher specific surface area, as shown in Table I, and require a higher binder content to yield a binder film as thick as with dense graded aggregate mixes.

For the 6 wt % asphalt – 12 wt % sulfur Thermopave formulation under consideration, assuming 20% extension of the asphalt with sulfur, the effective binder content is:

$$6\% + (6\% \times 0.20) = 7.2\%.$$

The remaining 10.8 wt % of sulfur performs as a mix filler. These large agglomerations of sulfur do not perform in the same way as conventional mineral fillers which are dispersed in asphalt hot-mix. The latter, at the concentrations typically used, generally improve the void filling capacity of the asphalt binder (or reduce the VMA) without effecting such dramatic changes to mix stability as does sulfur.

In summary, the laboratory studies demonstrated that aggregate shear during mixing yields 4-μ sulfur particle dispersions in asphalt, essential for long-term extension of asphalt in conventional hot-mixes. Increasing the sulfur content beyond the solubility/dispersion limit, believed to be approximately 20 wt % of asphalt, and use of substantially higher sulfur contents as in Thermopave resulted in coagulation of the excess sulfur and formation of large droplets which crystallized on cooling and ceased to perform as an effective part of the binder.

The mixing time required to attain a fine sulfur dispersion of uniform density depends upon mixer speed and design. The laboratory scale studies performed were judged inadequate for predicting the performance of large plant pugmills, and it was judged essential to verify the findings using several typical commercial hot-mix asphalt plants.

Sulfur Dispersion in Asphalt Plants. Sulfur-extended binder and Thermopave mixes were prepared in several batch type commercial hot-mix plants to verify the effectiveness of aggregate shear on sulfur dispersion during mixing in the pugmill. The influence of the following variables on the dispersion of sulfur was assessed.

(1) Sulfur-extended binder system—binder content, 6 wt % mix; binder composition, 4.5 wt % asphalt, 1.5 wt % sulfur; sulfur/asphalt weight ratio 0.33.

    (a) Aggregate—crushed stone ¼–½ in., 6–12 mm) was used as it has the least effective dispersive effect in the laboratory study.

    (b) Mixing plants—A range of batch-type models of various capacities was selected:

        (i) Barber–Greene

capacity:	5,000 lb	(2,268 kg)
batch size evaluated:	4,000 lb	(1,814 kg)

        (ii) Hethrington and Berner

capacity:	6,000 lb	(2,722 kg)
batch size evaluated:	4,000 lb	(1,814 kg)

        (iii) Cedarapids

capacity:	10,000 lb	(4,536 kg)
batch size evaluated:	7,000 lb	(3,175 kg)

    (c) Mixing times—liquid sulfur was added to the pugmill immediately after asphalt. Time required to add sulfur, 5 sec. Mixing times investigated (after sulfur addition), 15, 30, 45, and 60 sec.

(2) Thermopave mix (medium sand)—asphalt content, 6 wt %; sulfur content, 13 wt %; Plant, Cedarapids; capacity, 6000 lb (2722 kg) batch; batch size evaluated, 5000 lb (2268 kg); mixing time, 25 sec with sulfur.

Microscope slides of the sulfur–asphalt binders were prepared at the plant sites. For the sulfur-extended binder mixes, the slides were photographed within 2 hr of preparation. For the Thermopave mix, the slides were photographed two days after preparation.

Examination of thin sulfur-extended binder films showed good dispersion with most of the sulfur reduced to less than 4 μ in diameter after a 15-sec mixing time and only the odd particle of larger diameter present, as shown in Figure 2. After 30 sec of mixing with sulfur, binders from all of the plants showed, without exception, good dispersion and uniform density. Their typical appearance is shown in Figure 3. The binders mixed with sulfur for 45 and 60 sec were similar in appearance to Figure 3, demonstrating that mixing beyond an optimum time interval between 15 and 30 sec does not improve the quality of the dispersion.

This verifies that aggregate shear during mixing is effective in dispersing sulfur in commercial hot-mix plants. Addition of asphalt and sulfur to the pugmill simultaneously would permit the manufacture of sulfur-extended binder mixes at the usual mixing cycle time.

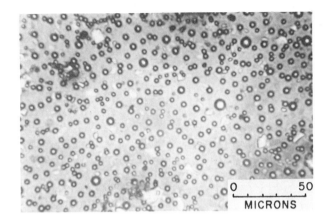

Figure 2. *Appearance of sulfur particles dispersed through-out asphalt phase of sulfur-extended binder mix after 15 sec mixing in Cedarapids pugmill. Batch size: 7000 lb.*

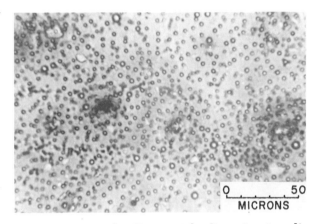

Figure 3. *Improved sulfur particle dispersion in sulfur-extended binder mix after 30 sec mixing in Cedarapids pugmill. Batch size: 7000 lb.*

Examination of thick sulfur-extended binder films, mixed with sulfur for 30, 45, and 60 sec, showed the presence of some larger sulfur particles. As the particle density on the slide is higher in thick films, this demonstrates that some coagulation of the sulfur particles takes place if the sulfur content is too great, relative to the amount of asphalt (sulfur/asphalt weight ratio, 0.33). Similar evidence of coagulation of particles was obtained in thick films taken from the laboratory Waring blender dispersion for a similar blend.

The dispersion in a thin sulfur–asphalt film from the Thermopave mix is shown in Figure 4. The presence of sulfur particles smaller than

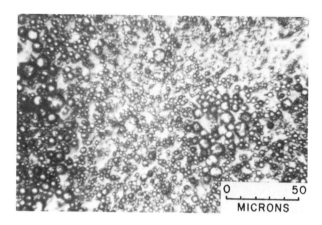

Figure 4. Thin sulfur–asphalt film from a Thermopave mix showing variety in sulfur particle size

4-μ diameter demonstrates that the shear during mixing effectively disperses the sulfur particles. The presence of the larger particles indicates that sulfur coagulation occurs rapidly because of the large amount of "excess" sulfur (sulfur/asphalt ratio, 2.0). The coagulation is significant in thick sulfur–asphalt films, as shown by the presence of large sulfur globules in Figure 5.

Table II. Effect of Asphalt Replacement

Mix No.	Binder Content (wt %)	Binder Composition		Marshall Stability after 3 Days	
		wt % Asph.	wt % Sulfur	(lb)	(N)
1	5	100	0	1947	(8661)
2	5	80	20	1540	(6850)
3	5	60	40	2017	(8972)
4	5	50	50	3420	(15212)

[a] Aggregate and asphalt B properties appear in the Appendix. Sulfur-extended

Table III. Effect of Asphalt Replacement

Mix No.	Binder Content (wt %)	Binder Composition		Marshall Stability after 3 Days	
		wt % Asph.	wt % Sulfur	(lb)	(N)
1	5.0	100	0	1947	(8661)
5	5.2	80	20	1180	(5249)
6	5.8	60	40	1766	(7856)
7	6.1	50	50	2858	(12713)

[a] Aggregate and asphalt B properties appear in the Appendix. Sulfur-extended

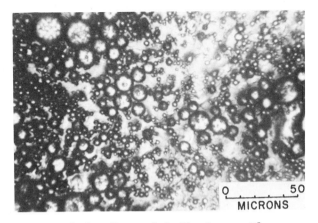

Figure 5.   Thick sulfur–asphalt film from a Thermopave
mix and large sulfur globules resulting from coagulation

Special Features of Sulfur-Extended Binder Mixes

A number of features distinguish sulfur-extended binder mixes from conventional hot-mix and must be considered in mix design. To illustrate this, samples have been prepared from a conventional asphalt aggregate mix and from a series of sulfur-extended binders. Information on the

by Sulfur on a Weight Basis[a]

Marshall Flow after 3 Days		Marshall Stability after 28 Days		Marshall Flow after 28 Days	
(0.01 in.)	*(mm)*	*(lb)*	*(N)*	*(0.01 in.)*	*(mm)*
8	(2.0)	2080	(9252)	8	(2.0)
7	(1.8)	2130	(9475)	8	(2.0)
6	(1.5)	3146	(13994)	7	(1.8)
8	(2.0)	3883	(17272)	7	(1.8)

binder prepared in Waring blender.

by Sulfur on a Volume Basis[a]

Marshall Flow after 3 Days		Marshall Stability after 28 Days		Marshall Flow after 28 Days	
(0.01 in.)	*(mm)*	*(lb)*	*(N)*	*(0.01 in.)*	*(mm)*
8	(2.0)	2080	(9252)	8	(2.0)
6	(1.5)	1636	(7273)	7	(1.8)
7	(1.8)	2540	(11298)	7	(1.8)
7	(1.8)	3366	(14973)	7	(1.8)

binder prepared in Waring blender.

design of these mixes and properties of the materials used appear in the Appendix.

The data in Tables II and III indicate that the Marshall stability of a mix is lowered initially when a low amount of asphalt is replaced by sulfur (i.e., 20%). This is attributed to softening of the binder by sulfur (14). After curing over a longer time (e.g., 28 days), the stability increases; the value for the conventional mix is surpassed by mix No. 2 but not by mix No. 5. A more detailed study of curing rates, shown in Figure 6, indicates that Marshall stability does not increase significantly after a curing period of approximately eight days.

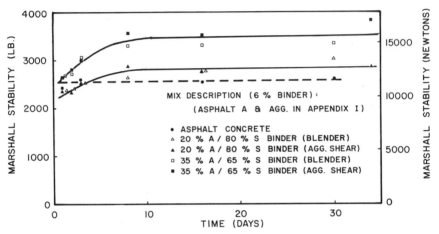

Figure 6. Effect of curing time on Marshall stability for mixes with sulfur-extended binders prepared in high-shear blender and by aggregate shear during mixing

Increasing the proportion of sulfur in the binder, as shown by mixes No. 3, 4, 6, and 7 in Tables II and III, increases Marshall stability. This is attributed to the excess particulate sulfur, dispersed throughout the asphalt phase, which performs similar to a mineral filler. The particular sulfur may be observed using a microscope and is visible as small yellowish specks along broken mix surfaces.

This increase in stability, however, is accomplished at the sacrifice of mix flexibility. The data in Table IV indicate that the flexural strain at break of mixes No. 3 and 4 is reduced significantly below that of mixes No. 1 and 2 when the sulfur concentration in the binder is raised to 40 and 50 wt %, without increasing the total binder content. Flexibility is a measure of mix susceptibility to brittle fracture and is an important mix design consideration in rigid materials to prevent premature pavement cracking.

Table IV.   Flexural Properties of Mixes[a]

Mix No.	Binder Content (wt %)	Binder Composition wt % Asph.	Binder Composition wt % Sulfur	Maximum Deflection (cm)	Maximum Tensile Strain	Stress at Break $(\times 10^6$ $N/m^2)$
1	5	100	0	0.50	0.0155	2.92
2	5	80	20	0.40	0.0121	1.92
3	5	60	40	0.15	0.0046	2.53
4	5	50	50	0.12	0.0038	2.59

[a] Flexure specimen loaded mid-span. Specimen, 30 mm $\times$ 20 mm $\times$ 200 mm; strain rate, 20 mm/min; test temp., 10°C; specimen curing time, 6–10 days. $e =$ $6ha/l^2$ where $e =$ maximum tensile strain (mm/mm), $h =$ specimen thickness, $a =$ deflection of specimen, and $l =$ length between supports.

It is therefore recommended that the sulfur concentration in the binder be maintained at a low level, sufficiently high to meet mix stability requirements but without lowering the standard of mix flexibility i.e., 20–30 wt %. Flexibility should not be confused with fatigue resistance, which

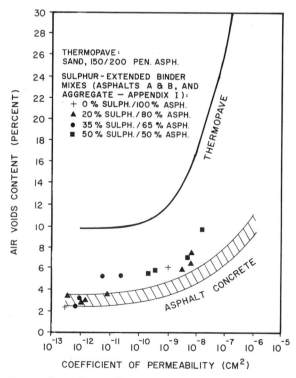

Figure 7.   Relation between air voids content and permeability for mixes

reflects the ability of a pavement to sustain repeated bending and which has been found to be satisfactory for a moderate range of sulfur-extended binder formulations (5).

The allowable sulfur concentration in the binder depends on the properties of the asphalt. For example, asphalts A and B (Appendix, Table A-I) exhibit significantly different viscosities at the Marshall test temperature of 60°C. This difference is reflected by differences in mix stability at similar asphalt contents, shown in the Appendix and in Figure 6, i.e., 11120 N and 5960 N for asphalt A and B, respectively, at a content of 6 wt %. Asphalt B yields high-stability mixes and is not as prone to softening by low sulfur concentrations in the binder, whereas asphalt A exhibits the reverse behavior.

The air-void content of asphalt concrete mixes affects mix permeability and is therefore very important from a pavement durability standpoint. The permeability of various sulfur-extended binder mixes, prepared at a variety of air-void contents, was measured with an air permeameter similar to that described in Ref. 16. The results, presented in Figure 7, indicate that the sulfur-extended binder mixes exhibit an air voids vs.

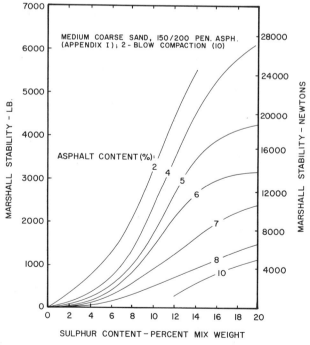

Figure 8. Influence of sulfur content on Marshall stability

Table V.   Influence of Mix Aging on Thermopave Stability[a]

Aging Time	Marshall Stability	% of Average 5-Yr Strength
5 days	1779	91
3 mo	1966	101
6 mo	1930	99
1 yr	1729	89
1½ yr	1895	97
2 yr	1877	96
3 yr	1808	93
4 yr	2006	103
5 yr	2201	113

[a] Mix: medium-coarse sand, 150–200 pen asphalt C, 6% A/12% S.

permeability relationship similar to asphalt concrete. As in conventional mixes, it is important therefore to maintain a low air-void content to maximize pavement durability. When replacing a portion of asphalt with sulfur, the higher specific gravity of sulfur should be considered, and the replacement of asphalt with sulfur should be on an equivalent volume basis to maintain existing standards for mix impermeability and durability.

Effect of Sulfur Concentration on Thermopave Properties

The use of higher amounts of sulfur, above a sulfur/asphalt weight ratio of 1.0, yields pourable mixes with a marked change in physical properties. For example, the sand mixes in Figure 8 exhibit negligible stability without sulfur; addition of sulfur permits mix designs to high stability levels at a variety of asphalt contents. Other mix properties are affected in a similar fashion.

Short-term studies indicate that Thermopave mixes generally attain their ultimate stability after a curing time between one to five days, depending upon the sulfur/asphalt ratio and sand type. The results of a long-term aging study of Marshall briquettes stored at room temperature are presented in Table V. The data indicate a random variation in stability values but no significant change in stability over a five-year period.

The predominant role of sulfur in Thermopave is to stabilize the mix while the asphalt contributes to mix flexibility. Control of these two properties provides flexibility to mix design, permitting attainment of a variety of distinctive mix characteristics. The mix design technology for Thermopave differs significantly from conventional mix design. Mix optimization to balance the various mix properties is involved. The limited scope of this chapter restricts their consideration in detail.

The properties of Thermopave are not as susceptible to change with temperature as are those of asphalt concrete. At warm temperatures the stiffness modulus (stress/strain ratio, t, T) of Thermopave is higher than for asphalt concrete. In the upper pavement service temperature range, when the asphalt stiffness is low, for example 1×10^6 N/m², the stiffness modulus of asphalt concrete is generally in the range from 3×10^8 to 9×10^8 N/m² whereas a typical sand Thermopave mix (6 wt % asphalt; 12 wt % sulfur) has a stiffness modulus of 2×10^9 N/m². This implies that Thermopave is less prone to deformation and has better load distribution capabilities at warm temperatures (T) or long-loading times (t).

Because of its higher rigidity at warm temperatures, sand Thermopave formulations are not as flexible as asphalt concrete mixes. A typical sand Thermopave mix (6 wt % asphalt; 12 wt % sulfur) exhibits a flexural strain at break of 0.004 cm/cm under the same test conditions as indicated in Table IV. Although this is below the strain values for asphalt concrete, lower flexibility in Thermopave can be tolerated as the tensile stresses and strains developed at the underside of the pavement are lower than for an asphalt pavement of equivalent thickness and subjected to the same loading. Performance of test pavements to date, some over six years old, have not indicated flexibility to be a problem as yet.

The fatigue life of well designed blow sand Thermopave mixes (e.g., over 50% finer than No. 100 mesh) is slightly below that of asphalt concrete. The fatigue life of mixes in which coarser sands are used, however, is comparable with that of asphalt concrete.

Naturally deposited sands generally have voids in the mineral aggregate (VMA) between 35–40 vol %. After optimizing the mix formulation, the air-voids content of the resultant mix may be in the range of 10–16 vol %. While this would be intolerable for asphalt concrete, it is quite acceptable for Thermopave. The relationship between air-voids content and mix permeability is shown in Figure 7 (10). Asphalt concretes with air-void contents of 6%, for example, exhibit a coefficient of air permeability of approximately 10^{-8} cm². In Thermopave the equivalent degree of mix impermeability is attained at an air-void content of approximately 16%, indicating that the material is relatively impervious, and its durability is not expected to be impaired despite a high air-void content (i.e., below 16%).

Thermopave generally has a density 10–14% lower than conventional asphalt concrete. Additional laboratory studies and successful performance of test pavements have encouraged us to revise our initial conservative position published in 1971 (8) and to recommend Thermopave structural equivalency on par with asphalt concrete. Because of its higher yield, a given tonnage of Thermopave will build a 10–14% longer pavement than the same weight of asphalt concrete.

Conclusions

Laboratory studies of sulfur-extended binder mixes demonstrated that sulfur is readily dispersed in asphalt by aggregate shear during mixing, irregardless of the mode in which the constituents are combined and the nature of the aggregates. This was verified in commercial Barber-Greene, Hethrington and Berner, and Cedarapids hot-mix plants with batch sizes as large as 7000 lb (3175 kg). The optimum mixing time to attain fine sulfur dispersions of uniform density with diameter less than 4 μ, the governing particle size criterion, was 15–30 sec after sulfur addition. This is either less than or equal to the wet-mix time used at most asphalt plants, indicating that production capacity is not reduced if both asphalt and sulfur are introduced into the pugmill simultaneously. The technique of microscopic examination implemented can be used readily to establish the minimum mixing time required to attain dispersion of sulfur for any particular plant.

Thermopave mixes use substantially higher sulfur contents. Microscopic examination of mixes prepared in the laboratory and in a commercial Cedarapids hot-mix plant verified that some of the sulfur is finely dispersed, therefore contributing to the dilution of the asphalt, but most of the sulfur agglomerates readily to perform as a mix stabilizer.

The studies illustrate that the two sulfur–asphalt systems are complementary. The primary objective for using the sulfur-extended binder system is to use sulfur as a diluent to conserve asphalt used in conventional hot mixes. No asphalt saving is realized using Thermopave. The unique features of this process are that it permits the manufacture of high-quality paving materials using inexpensive, poorly graded sands and yields pourable mixes, similar to cement concrete.

Based on studies performed, some guidelines have been formulated for the design of sulfur-extended binder mixes. These include recommendations for limiting sulfur content to maintain existing mix flexibility and the replacement of asphalt with sulfur on an equivalent volume basis to maintain the existing standard of mix durability.

The mix design technology for Thermopave is more involved than for asphalt concrete. The main Thermopave property considerations have been outlined for optimizing mix formulations. Studies to date indicate that properly designed Thermopave mixes may be used on par with asphalt concrete on an inch-for-inch basis. Since the bulk density of Thermopave is lower, it yields a 10–14% greater length of pavement than the equivalent tonnage of asphalt concrete.

188 NEW USES OF SULFUR—II

Appendix

Table A-I. Mix Aggregate Gradation (Percent Passing)

ASTM Sieve Size	Sulf-Extended Binder Agg.	Sand Used in Thermopave
½ in.	100	
⅜ in.	82–86	
No. 4	52–57	
8	41–50	100
16	34–42	98
30	22–30	81
50	10–12	49
100	5	14
200	2.5–3.0	2.4

Table A-II. Asphalt Properties[a]

	A	B	C
Penetration at 25°C	90	105.5	170
Penetration at 5°C	12	10.5	
Viscosity at 135°C, cst.	430	311	232
Viscosity at 60°C, poise	2077	1164.8	576

Table A-III. Marshall Mix Design Properties[a]

Mix Property	Asphalt Content (wt %)			
	3	4	5	6
Marshall stability (lb)	1160	1593	1853	1340
(N)	5160	7086	8243	5961
Marshall flow (0.01 in.)	6	7	8	8
(mm)	1.5	1.8	2.0	2.0
Bulk density (g/cc)	2.292	2.297	2.323	2.358
Volume air voids (%)	8.5	6.9	4.4	1.6

[a] Asphalt B used in mix design. Optimum asphalt content, 5.0%.

Literature Cited

1. Sullivan, T. A., McBee, W. C., Rasmussen, K. L., "Studies of Sand–Sulphur–Asphalt Paving Materials," *U.S. Bur. Mines Rep. Invest.* (1975) **8087**.
2. Societe Nationale des Petroles d'Aquitaine, "Properties of Sulphur Bitumen Binders," Documentation, VII IRF World Meeting, 1973.
3. Kennepohl, G. J. A., Logan, A., Bean, D. C., "Sulfur–Asphalt Binders in Paving Mixes," *Can. Sulfur Symp. (Pap.)* (1974).
4. Kennepohl, G. J. A., Logan, A., Bean, D. C., " 'Conventional' Paving Mixes with Sulphur–Asphalt Binders," *Proc. Assoc. Asphalt Paving Technol.* (1975) **44**.
5. Kennedy, T. W., et al., "An Engineering Evaluation of Sulphur–Asphalt Mixtures," Transportation Research Board Record (1977), in press.

6. Pronk, F. E., et al., "Sulphur Modified Asphaltic Concrete," *Proc. Can. Tech. Asphalt Assoc.* (1975) **XX**.
7. Lee, D. Y., "Modification of Asphalt and Asphalt Paving Mixtures by Sulfur Additives," *Ind. Eng. Chem., Prod. Res. Dev.* (1975) **14** (3).
8. Hammond, R., Deme, I., McManus, D., "The Use of Sand–Asphalt–Sulphur Mixes for Road Base and Surface Applications," *Proc. Can. Tech. Asphalt Assoc.* (1971).
9. Deme, I., "Basic Properties of Sand–Asphalt–Sulphur Mixes," Documentation VII, IRF World Meeting, 1973.
10. Deme, I., "Processing of Sand–Asphalt–Sulphur Mixes," *Proc. Assoc. Asphalt Paving Technol.* (1974).
11. Burgess, R. A., Deme, I., "The Development of the Use of Sulphur Asphalt Paving Mixes," *Adv. Chem. Ser.* (1974) **140**, 85.
12. Shane, G., Burgess, R. A., "The Thermopave Process," Proc. Symposium on New Uses for Sulphur, Madrid, 1976.
13. Bencowitz, I., Boe, E. S., "Effect of Sulphur Upon Some of the Properties of Asphalts," Proc. ASTM (1938) **38**, Part 2.
14. Quarles van Ufford, J. J., Vlugter, J. C., "Schwefel und Bitumen, I," *Brenst. Chem.* (1962) **43**.
15. Petrossi, U., Bocca, P. L., Pacor, P., "Restrictions and Technological Properties of Sulfur-Treated Asphalt," *Ind. Eng. Chem. Prod. Res. Dev.* (1972) **11** (2).
16. Davies, J. R., Walker, R. N., "An Investigation on the Permeability of Asphalt Mixes," Ontario Ministry of Transportation and Communications, Research Report 145, 1969.

RECEIVED April 22, 1977.

11

Prediction of In-Service Performance of Sulfur–Asphalt Pavements

D. SAYLAK, R. L. LYTTON, B. M. GALLAWAY, and D. E. PICKETT

Texas Transportation Institute, Texas A&M University,
College Station, TX 77843

Three approaches to incorporating sulfur into an asphaltic concrete mix as well as a unique mix design method using nonlinear programming have been developed. The predicted maximum resilient moduli of optimum mixtures using the three approaches are compared by this mix design program. Using a linear viscoelastic layered pavement analysis program (VESYS IIM), the performance of four different sulfur–asphalt concrete pavements including a recycled mix with a conventional asphaltic concrete pavement are compared based upon predicted fatigue life, rutting potential, slope variance, and serviceability index of two pavement structures in three different climates for each type of surfacing material. The status of domestic field test projects currently in post construction evaluation as well as those in the planning stage is discussed.

The present inventory of pollution abatement sulfur is increasing by about 4 million tons per year, and by 1980 the supply could begin to exceed the demand (*1*). On the basis of this outlook, industry, government, and university groups have initiated a considerable amount of research and development to develop new uses for sulfur.

One of the most promising outlets for sulfur is highway construction. Interest in this application is being stimulated by two factors—the decreasing availability or total absence of quality aggregates in a number of regions around the country and the current increase in cost and projected construction and energy demands for asphalt cement. The demand for construction aggregates is increasing at a time when sources near urban and other high-use areas are being depleted. Furthermore, there are a number of locations such as the Gulf Coast states where aggregates must be transported long distances to accommodate the market. At the same time, the current energy squeeze has produced considerable uncertainty about the future availability of asphalt cement for road building purposes and, as a consequence, has caused bid prices to more than double over the past three years (2).

Sulfur's unique properties permit it to be utilized as an aggregate, as an integral part of the binder, or as both. As a result, sulfur-producing industries and research agencies in Canada, Western Europe, and the United States have stepped up research and development on the use of sulfur in highway pavement construction (3, 4, 5, 6, 7). Some of these activities have been related directly to the utilization of sulfur in a manner which will permit the use of lower quality, locally available aggregates. This approach was pioneered by Shell Canada, Ltd. (1) and resulted in a patented sand–asphalt sulfur mix called Thermopave (8). The Texas Transportation Institute (TTI) (9) is currently extending the technology developed by Shell for application in the U. S.

There is considerable research activity in the U. S. (2, 10, 11, 12), Canada (4), and Europe (5) on the partial replacement of the asphalt cement as the binder in asphaltic concrete. Concurrently, there are processes in which sulfur has been substituted successfully for 30–50 wt % (w/o) of the asphalt cement in otherwise conventional bituminous mixtures. These processes have been both developed and demonstrated independently by the Société Nationale des Pétroles d'Aquitaine (SNPA) in France (5) and by Gulf Oil Canada (4) using techniques and equipment which are proprietary to each. During autumn 1975, TTI, under the sponsorship of SNPA, The Sulphur Institute, FHWA, and Texas State Department of Highways and Public Transportation conducted a series of verification studies using the SNPA process which culminated in a successful field demonstration on U.S. 67 near Lufkin, Tex. (12). The binder in this case was a sulfur–asphalt emulsion in which 30 wt % (w/o) of the binder was sulfur. Gulf Canada, in 1972, placed a pavement mixture with 50 w/o sulfur in the binder (4). The Sulfur Development Institute of Canada (SDIC) is developing a paving mixture using sulfur-extended asphalt binders with complementary processes developed by Gulf Canada and F. E. Pronk of Highway Associates, Calgary, Alberta (13). The latter process uses a proprietary additive along with an in-line blending unit, whereas Gulf (and SNPA) utilizes a high-shear rate emulsifying unit which can produce a sulfur–asphalt emulsion binder at a rate of 70 tons per hr. Road tests using the SNPA process have been carried out in Spain and Finland during 1976 with additional demonstrations under consideration in Belgium and Germany.

The Bureau of Mines has produced quality sulfur–asphalt paving materials which are blended by the shearing actions generated in the pugmill (2). This eliminates the need for the separate high shear rate blender. During January 1977, the Bureau, in conjunction with the Nevada Highway Department, constructed a sulfur–asphalt field test pavement using a binder composed of 30 w/o sulfur. This pavement is now under post-construction evaluation.

An additional approach which will result in the conservation of asphalt, which is under investigation by both the Bureau of Mines and TTI, involves using sulfur in recycling old bituminous pavements (2, 14). Sulfur can reduce the viscosity and hence improve the workability of age-hardened asphalt pavement material and can increase the stiffness of the recycled mix when it cools.

This chapter discusses current research on the use of sulfur in recycled asphaltic concrete pavements. In addition, it describes the results of laboratory tests and theoretical predictions using the latest linear viscoelastic layered pavement analysis methods (15, 16) to compare the performance of various sulfur–asphalt concrete pavements with conventional asphalt concrete pavements in a variety of climates. The relationship between pavement distress and performance used in the computer program was established at the AASHTO road test (17). Finally, the results of domestic field tests of sulfur–asphalt pavements are presented along with a discussion of future trends for the utilization of sulfur in the construction of highway pavement materials.

Sulfur–Asphalt Pavement Analysis and Design

There are four major considerations in designing a sulfur–asphalt pavement—mix design, construction, performance, and economics. None of these are unique to sulfur–asphalt, but they must all be considered explicitly in order to show under which conditions the new sulfur–asphalt paving material may be expected to be superior to the widely used and accepted asphalt concrete pavement.

Mix design, performance, and economics have been considered traditionally in pavement design in a largely empirical or intuitive way which does not allow simple comparisons to be made between candidate materials such as concrete and asphalt. A more efficient and consistent means of considering them is warranted because of the current need to use the most economical materials in a rapidly changing market situation.

Mix design is the choice of the best combination of aggregate, binder, and compaction effort to produce desirable pavement material properties. The choice has been made normally based on the results of simple indicator and control tests such as the Marshall stability, Marshall flow, Hveem stability, resilient modulus, and indirect tension tests.

Construction considerations influence directly the economics of any paving procedure. Introduction of a new paving material or method will probably encounter certain obstacles, some of which may be physical and some psychological. In order to overcome these barriers, it is important that construction procedures suggested to the paving industry be based upon well documented filed trials. Of particular interest in evaluating construction techniques for sulfur–asphalt pavements is the utilization of available equipment for mixing, placing, and compacting the mixture. Other items which must be considered include special handling techniques, temperature control and workability of the mixture, curing of the pavement, weather limitations on placing the mixture, and the ease of quality control and testing. Safety and environmental considerations are also important. These items are treated in detail later in this chapter.

The performance of a pavement is the history of its level of service over its lifetime to the traveling public. The pavement that remains smooth and uncracked with few patches after 10 years is generally thought to perform well. Consistent measures of its present level of service are in standard use and are considered in some detail later.

The most economical pavement is the one which has the least overall total cost, including construction cost, as well as all future maintenance, rehabilitation, and user time delay costs. Minor variations in pavement thickness or material properties can sometimes make significant differences in the total cost of a pavement.

An Efficient Method of Mix Design. The best combination of aggregate properties, sulfur and asphalt quantities, and compaction effort may be considerably different in a sulfur–asphalt mix than in conventional asphalt concrete. Since there is little available field or laboratory experience to borrow upon, an experimental design procedure may be adopted to use laboratory experimental data most efficiently in determining the best mix.

The relevant properties of the constituents of sulfur–asphalt and the numerical scale that was developed for each of these independent variables are listed below. The ranges of the independent variables for the materials and processes used in the screening tests are presented in Table I.

1. Aggregate roundness can vary between 0.0 for a highly angular particle to 1.0 for a highly round particle.

2. Aggregate sphericity is an index of how closely a particle represents a sphere and can vary between 0 for a highly nonspherical particle to 1 for a highly spherical particle.

3. Aggregate surface roughness is a measure of the surface texture of the particle and is based upon an arbitrary scale ranging between 1 for very rough and 4 for polished texture.

4. Aggregate size and size distribution (gradation).

Table I. Range of the Independent Variables for the AAS System

Independent Variable	Upper Bound	Lower Bound
Aggregate roundness	0.78	0.10
Aggregate sphericity	0.87	0.70
Aggregate surface roughness	3.5	1.2
Aggregate size and gradation number (SGN) [a]	977	2
Asphalt cement viscosity	1	0.1
Binder content as v/o VMA [b]	99.9	1.0
Sulfur content as v/o binder [b]	81.8	50.0
Moisture condition	1	0.1
Compaction effort	1	0.1

[a] $SGN = \dfrac{D_{85} - D_{50}}{(D_{20})^2}$

[b] v/o designates percent by volume.

5. Asphalt cement viscosity varies between 0.1 for an AC5 and 1.0 for AC20 asphalt cements.

6. Binder (sulfur and asphalt) content.

7. Sulfur content

8. Moisture condition of 0.1 indicates the specimen was tested in a dry condition whereas 1.0 indicates the specimen has been vacuum saturated prior to testing.

9. Compaction effort variables of 0.1 and 1.0 indicate compaction efforts of 2 and 75 blows per face, respectively.

A series of laboratory screening tests were performed to evaluate the influence of the nine independent variables. To reduce the required number of screening tests, a fractional factorial experimental design was made using high and low values of the nine variables listed. A one-eighth factorial was sufficient to determine the combination of variables needed to evaluate their influence on the mixture.

Three individual systems of mixing sequences were used to incorporate the constituents in mixtures tested in the screening test phase. The systems are termed aggregate–asphalt–sulfur (AAS), aggregate–emulsion (AE), and aggregate–emulsion–sulfur (AES). The sequence of letters corresponds to the sequence with which the individual constituents were introduced into the mixer. For example, the first "A" in the AAS system indicates aggregate, the second "A" indicates introduction of asphalt cement which was mixed with the aggregate, and the "S" indicates sulfur which was then introduced and mixed during a second wet mix cycle. In the AE and AES systems, the "E" designates the emulsion produced by blending heated asphalt cement and sulfur in a high-speed colloid mill. The AAS system is similar to the Shell process (18, 19, 20) whereas the AE system is similar to the SNPA process (5). The AES system is a combination of the SNPA and Shell processes. These three systems were tested to evaluate the effects of varying the sequence of introduction of the individual components into the mixture.

For the screening test phase, three replicate specimens of each combination of independent variables were made, and each of the indicator or control tests was run on the specimens of sulfur–asphalt concrete. These indicator tests (dependent variables) included bulk specific gravity, air voids, voids in the mineral aggregate (VMA), resilient modulus, Hveem stability, Marshall stability, and Marshall flow. Table II presents the range of dependent variables determined during the screening test for the AAS system mixtures.

Mix design is as much an art as a science because interactions among the variables produce effects which must always be considered. The effects of these variables and their interactions are considered explicitly by developing quadratic regression equations of the form:

$$y = a_o + \sum_{i=1}^{9} a_i x_i + \sum_{i,j=1}^{9} a_{ij} x_i x_j$$

where the x_i are the nine independent variables, and y is the dependent variable resulting from an indicator or control test. In all cases, the regression equations had coefficients of determination (R^2) above 0.90. From Table II, it is obvious that very large variations in material properties can be achieved by varying the mix design.

As would be expected, significantly different regression equations were developed for the AE, the AAS, and the AES processes of combining the sufur–asphalt binder with the aggregate. The task of designing the mix for each of these systems now becomes a search for the best combinations. It is this search that is normally done by trial and error and intuition, but it may be accelerated by the use of nonlinear programming techniques (*21*).

The objective function to be maximized is the value of the indicator test that is considered to be most important. The constraints which must

Table II.   Range of the Dependent Variables for the AAS System

Dependent Variable	Upper Bound	Lower Bound
Bulk specific gravity	2.46	1.13
Voids in the mineral agg. (VMA) (%)	49.0	16.0
Air voids (%) .	34.3	0.0
Resilient modulus (psi)	4.73×10^6	0.06×10^6
Hveem stability	88	1
Marshall stability (lb)	29,512	0
Marshall flow (.01 in.)	25	4

be satisfied are either the minimum specified values of the other indicator tests that must be satisfied by standard asphalt mixes or the normal ranges of the independent variables that are encountered in realistic mix designs.

There are several advantages to this method of mix design using these regression equations in a nonlinear programming scheme. An example of one major advantage is the case where only poor aggregate is available for construction. Using these equations, it will be possible to determine the best combination of asphalt content, asphalt viscosity, and sulfur content to maximize a desirable mix criterion such as Marshall stability or resilient modulus without having to run extensive tests to determine the best ranges of these independent variables by trial and error.

It has been found in the course of these screening tests and the associated analysis that the resilient modulus is one of the most reliable indicators of the in-service performance of a pavement. A high resilient modulus is associated with low rutting and roughness and a higher serviceability index, but also a higher incidence of fatigue cracking.

Using the resilient modulus as the best indicator test, it was found that the AAS process produces the best mix with all qualities of aggregate. Efforts are being made to verify this tentative conclusion by performing various characterization tests on mixtures made by the AAS, AE, and AES processes in a laboratory testing program currently being conducted at Texas Transportation Institute. In all cases, the sulfur–asphalt pavement will have a resilient modulus that is substantially higher than the comparable mix with asphalt cement binder alone.

Mechanical Characterization of Sulfur–Asphalt. The serviceable life of a pavement comes to an end when the distress it suffers from traffic and climatic stresses reduces significantly either the structural capacity or riding quality of the pavement below an acceptable minimum. Consequently, the material properties of most interest to pavement designers are those which permit the prediction of the various forms of distress—resilient modulus, fatigue, creep, time–temperature shift, rutting parameters, and thermal coefficient of expansion. These material properties are determined from resilient modulus tests, flexure fatigue tests, creep tests, permanent deformation tests, and thermal expansion tests.

Characterization tests reported here were performed on combinations of materials needed to evaluate the predicted performance of typical pavements made from conventional asphaltic concrete with limestone aggregate, sulfur–asphalt concrete with limestone aggregate, sulfur–asphalt concrete with beach sand aggregate, and recycled asphaltic concrete pavement with sulfur added during the recycling process. The data obtained from these characterization tests were used as input for the VESYS IIM computer program (15) to predict the performance of the selected pavements. The crushed limestone aggregate used in the mixes had an SGN (size and gradation number) of 977 which complies with the requirements of the Asphalt Institute type IVb gradation. The beach sand aggregate was a uniformly graded fine said which had an SGN of 64. The aggregate gradation of the recycled mix obtained from Boulder City, Nev. conformed to Asphalt Institute Type IVb gradation. In addition, sulfur–asphalt mixtures made with volcanic aggregate which had an SGN of 761 were tested in flexural fatigue and for resilient modulus.

Sulfur contents may be expressed either on a weight or volumetric basis. The binder consisting of asphalt cement and sulfur may be expressed also as a percentage of the VMA, and the amount of sulfur may be expressed as vol % of the binder. The symbol used for wt % is w/o, whereas that used for vol % is v/o. Whichever basis of expression is used, it should be remembered that the specific gravity of sulfur is approximately twice that of asphalt cement, and therefore will occupy about half the space of an equivalent weight of asphalt cement. Asphalt cement and sulfur contents reported here were based upon wt % of total mix.

Table III. Resilient Modulus Results of Sulfur–Asphalt Concrete Mixes

Type Aggregate in Mix	Asphalt Content (w/o)	Sulfur Content (w/o)	Resilient Modulus $\times 10^6$ (psi)
Limestone	6.0	0.0	0.60
Limestone	5.2	1.8	1.47
Limestone	4.7	2.9	1.64
Limestone	4.2	4.1	2.36
Volcanic	6.0	0.0	0.65
Volcanic	5.1	2.1	0.98
Volcanic	4.4	3.4	1.54
Volcanic	3.8	4.7	2.04
Beach sand	6.0	13.5	0.58
Recycled boulder mix	5.5	1.25	1.95

Resilient Modulus. The Schmidt test device (22) was used to measure the resilient modulus of all samples at 68°F. A load pulse time of 0.1 sec and a rest time between pulses of 2.9 sec was used in this procedure. The results of these tests are presented in Table III.

Resilient modulus results of the limestone and volcanic aggregate mixes are shown grapically in Figure 1. As the sulfur content increases,

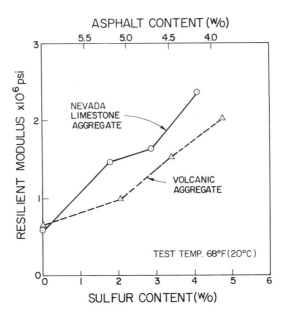

Figure 1. *Variation of resilient modulus with sulfur and asphalt content*

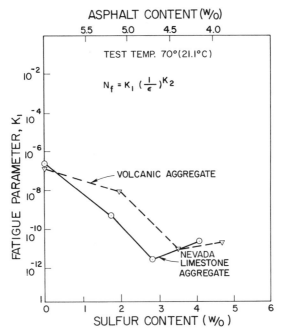

Figure 2. *Variation of K_1 with sulfur and asphalt content*

the resilient modulus also increases. At a sulfur content of about 4 w/o of total mix, the resilient modulus is about three times that of conventional asphaltic concrete.

Flexural Fatigue Tests. The procedure for the flexure fatigue test is described in detail in the Federal Highway Administration Users Manual for the VESYS IIM computer program (*15*). The fatigue tests are third point, upward loading, stress-controlled tests performed on beam specimens as described in the reference. Since strain continually increases during these tests, the strain reported here is the initial strain obtained by dividing the stress amplitude by the resilient modulus. The beams were tested in a half-wave sine loading at a frequency of 100 cycles per min (1.67 Hz). The equation used in VESYS IIM to relate initial strain and number of cycles to failure is:

$$N_f = K_1 \left(\frac{1}{\epsilon}\right)^{K_2} \tag{1}$$

where N_f is the number of repetitions to failure, and ϵ is the initial strain.

Both K_1 and K_2 are fatigue parameters which vary with the composition of the asphalt mixture. Figures 2 and 3 present the variation of K_1

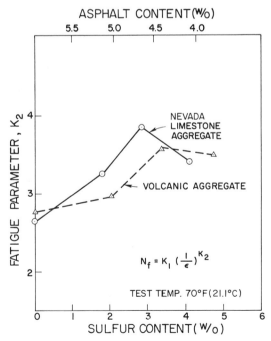

Figure 3. Variation of K_2 with sulfur and asphalt content

and K_2 respectively, with sulfur content for mixtures made with the limestone and volcanic aggregates. The value of K_1 decreases as sulfur content increases. Figure 3 shows that K_2 increases with sulfur content and is virtually the mirror image of the logarithmic plot of K_1. The shapes of the K_1 and K_2 curves for the volcanic and limestone aggregate indicate the similar fatigue behavior of mixes made with these two aggregates. As listed subsequently in Table IV, the values of K_1 used to evaluate the four different mixtures ranged between 25×10^{-7} and 2.5×10^{-20}, whereas the values for K_2 ranged between 2.55 and 5.67.

The fatigue performance of these materials in a pavement cannot be inferred directly from these values of K_1 and K_2, for the actual fatigue behavior will depend upon such variables as thickness and stiffness and size of tire footprints. In addition, it is well known that laboratory fatigue data does not represent accurately the actual fatigue strength of pavements. A given material in a pavement can take 10–100 times as much strain under seal traffic loads as it can when loaded under laboratory conditions to the same number of load cycles.

Creep Tests. All creep tests were run according to the procedure described in the Federal Highway Administration VESYS IIM Users Manual (15). The tests were performed on cylindrical samples 4 in. (10 cm) in diameter and 8 in. (20 cm) high. Creep tests of varying duration were performed at time spans between 0.1 and 1000 sec. There are a variety of ways to represent the creep compliance as a function of time, but perhaps one of the most useful is the power law form:

$$D(t) = D_o + D_1 t^n \qquad (2)$$

where D_o is the glassy compliance, D_1 is the intercept of the straight line portion of the creep compliance curve on log–log paper with the 1-sec time line, and n is the slope of the logarithmic creep compliance curve.

For the VESYS IIM program, creep compliance values were calculated from:

$$D(t) = \frac{\epsilon(t)}{\sigma_o} \qquad (3)$$

Table IV. Surface

Type Mix	Asphalt Content (w/o)		Sulfur Content (w/o)
Asp. conc. w/limestone agg.	6.0		0.0
Sulfur–asp. conc. w/limestone agg.	3.9		4.1
Sulfur–asp. conc. w/beach sand	6.0		13.5
Recycled boulder mix w/sulfur	5.0	(orig.)	1.5
	1.25	(added)	

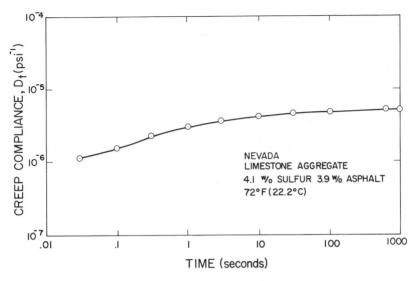

Figure 4.   Typical creep compliance curve

where $\epsilon(t)$ is the axial strain at time, t, and σ_o is the applied vertical stress. A typical creep compliance curve for one of the sulfur–asphalt concrete mixes made with limestone aggregate is shown in Figure 4. This curve comes from a 1000-sec creep test. Similar curves were obtained for each of the pavement mixtures evaluated in this program.

Time–Temperature Shift Characteristics. Temperature is a major environmental influence on viscoelastic pavement response. The VESYS IIM program can handle material properties as a function of temperature variations. The computer input command BETA relates the time–temperature shift factor, a_t, to the temperature variable for the pavement materials. This relationship is given by:

$$\text{Log } a_t = \beta(T_o - T) \tag{4}$$

Layer Properties

K_1	K_2	$BETA$ β	$ALPHA$ α	GNU μ
2.50×10^{-7}	2.65	0.113	0.650	0.145
3.00×10^{-11}	3.42	0.0738	0.800	0.098
2.00×10^{-9}	3.94	0.10	0.674	0.0778
2.51×10^{-20}	5.67	0.05	0.562	0.0225

where β is the value input for BETA, T_0 is the reference temperature, T is the temperature variable, and a_t is the time–temperature shift factor for a temperature T. The time–temperature shift factor is determined by:

$$a_t = \frac{t_T}{t_{T_0}} \tag{5}$$

where t_T is the time to obtain a given value of a material property at temperature T, and t_{T_0} is the time to obtain the same value of the material property at the reference temperature, T_0.

The values of BETA are determined by performing creep tests at various temperatures on the pavement mixtures. BETA values used in this study are reported in Table IV.

Repeated Load Triaxial Tests for Permanent Strain Evaluation. In accordance with the procedure in the VESYS IIM Users Manual, the same sample used in the creep tests was also used to measure permanent strain, an important indicator of the rutting potential of a pavement. Repetitive loads of the haversine type were applied to the specimen at a magnitude equal to the applied vertical stress used in the creep tests. The load duration was 0.1 sec followed by a rest period of 0.9 sec.

Data from the repeated load triaxial test are used to calculate permanent strain which can be plotted vs. load repetitions on log–log scale. Figure 5 shows a typical plot of accumulated axial strain vs. number of load repetitions. The confining stress was zero for all triaxial tests.

Each curve has its own intercept, I, and slope, S, which are used to calculate input data for the VESYS IIM permanent deformation subprogram. The equation of the curve is:

$$\epsilon_a = I N^S \tag{6}$$

where ϵ_a is accumulated permanent strain, I is intercept of the curve with the vertical axis at one repetition, N is number of repetitions, and S is the absolute slope of the curve.

The incremental amount by which the strain increases with each load repetition is:

$$\frac{d\epsilon_a}{dN}(N) = ISN^{S-1} = \Delta\epsilon_a \tag{7}$$

The fractional increase of the total strain, $F(N)$, that is permanent with each load repetition is:

$$F(N) = \frac{\Delta\epsilon_a}{\epsilon_r + \Delta\epsilon_a} \tag{8}$$

If it is assumed that the resilient strain, ϵ_r, is large compared with the increase of the permanent strain with each load repetition, then the following approximation can be made:

$$F(N) = \frac{\Delta\epsilon_a}{\epsilon_r} = \frac{IS}{\epsilon_r} N^S - 1 \qquad (9)$$

Equation 9 may be shown as follows:

$$F(N) = \mu N^{-\alpha} \qquad (10)$$

where $\mu = IS/\epsilon_r$, and $\alpha = 1 - S$. The constants α and μ are used as input data for the VESYS IIM program. Values of the constants used for this

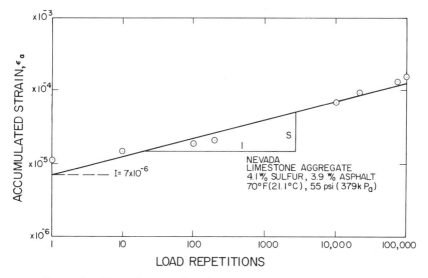

Figure 5. *Typical accumulated strain curve from triaxial test data*

study are reported in Table IV. When the power α equals 0, the percent of incremental strain increase with each load remains constant at a value of μ. When the power α is positive, the fraction $F(N)$ gets smaller with each load application. A negative α means that there is a progressive increase in strain with each additional loading.

Assumed Pavement Parameters. Two typical pavements were selected for this study. One of these consisted of a 2-in. thick surface course on top of an 8-in. thick base course, and the other pavement has a 3-in. thick surface course land over a 10-in. thick base course. These two pavements were assumed to be placed in three different climates whose temperature variations are listed in Table V.

Table V. Assumed Climatic Temperature Variations (°F)ᵃ

Month	Cool	Moderate	Warm
1	11.5	21.5	41.5
2	13.5	23.5	43.5
3	15.5	25.5	45.5
4	35.5	45.5	65.5
5	50.5	60.5	80.5
6	59.0	69.0	89.0
7	63.0	73.0	93.0
8	65.0	75.0	95.0
9	61.0	71.0	91.0
10	57.5	67.5	87.5
11	34.5	44.5	64.5
12	29.0	39.0	59.0

ᵃ °C = 0.56°F − 17.78.

Traffic was assumed to increase from 3000 heavy vehicle axles per day to 8000 heavy vehicle axles per day within 20 years. The intensity of loading was about 74 psi over an area with a radius of 6.4 in. for 0.1 sec. These traffic conditions are considered to simualte a heavily traveled roadway and were input in the VESYS IIM program as required.

The subgrade used in the analysis was assumed to be a clay material (A-6 classification) with a moisture content of about 23% and an elastic modulus of 15,000 psi. The base course used in the analysis was assumed to be a dense graded crushed aggregate with an elastic modulus of 60,000 psi. These two materials were assumed to be elastic and are typical of those used in construction of many pavements throughout the U.S.

The surface courses for the selected pavements were assumed to consist of either conventional asphaltic concrete with limestone aggregate, sulfur–asphalt concrete with beach sand aggregate, or recycled Boulder asphaltic concrete with sulfur. Material properties used as VESYS IIM input data for these four surface layers are listed in Table IV.

Prediction of Performance. The VESYS IIM program was used to assess the structural integrity of the two assumed pavements with the various surface materials. This program computes pavement distress in terms of rutting, roughness, and crack damage. These distress indicators are then used in a distress performance relationship to predict the serviceability history of the selected pavement.

Rut depth measures the depression in the pavement surface created by traffic in the wheel path. Figure 6 presents the predicted increase of rut depth with time for the pavement with the 3-in. thick surface layer in the warm climate. The data trend was similar for the other combinations of surface thicknesses and climates, although the actual values of predicted rut depth varied. Generally, the order of the surfacing materials showing progressively increasing rut depths was sulfur–asphalt concrete made with limestone aggregate, recycled Boulder mix, beach sand mix, and conventional asphaltic concrete. The one exception to this general data trend was the 2-in. thick surfaced pavement in the moderate and cool climates where the rutting of the beach sand mixture was greater than that of the conventional asphaltic concrete for the 2-in. thick surface layer in the moderate and cool climates. The data indicate that the addition of sulfur decreases significantly the predicted rut depth. The expected rut depths for a particular pavement section increased with increasing climatic temperatures.

Roughness of the pavement surface can be expressed in terms of the variance of the slope measurements of deformation taken along the longitudinal profile of the roadway in accordance with the American Association of State Highway and Transportation Officials (AASHTO) definition. In the VESYS IIM program, the variance of rut depth is computed in the rut depth model and is passed automatically to the roughness model. The difference in rut depth is assumed to occur because of material variability of each layer. The roughness output is in the

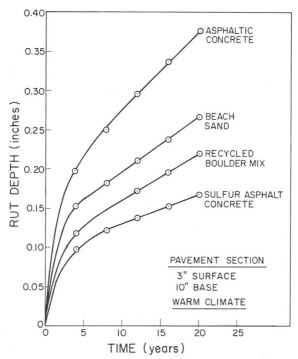

Figure 6. Comparison of predicted rut depth for sulfur–asphalt and asphaltic concrete mixtures

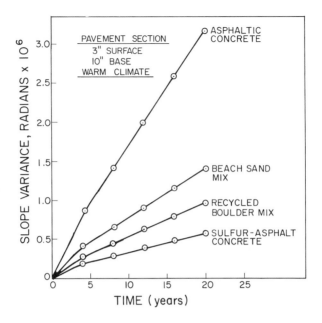

*Figure 7. Comparison of slope variance for 3-in.
thick sulfur–asphalt and asphalt concrete pavements*

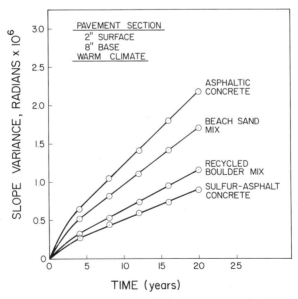

*Figure 8. Comparison of slope variance for 2-in.
thick sulfur–asphalt and asphalt concrete pavements*

form of slope variance. Figures 7 and 8 present the predicted slope variance as a function of time in the warm climate for the 3-in. thick and 2-in. thick pavement sections, respectively. In comparing the two figures, the slope variances of the 3-in. thick pavements are expected to be less than for the 2-in. thick pavements constructed of all the trial surfacing materials except the conventional asphaltic concrete. The results indicate that the 3-in. thick pavement made of conventional asphaltic concrete would be rougher than the 2-in. thick pavement of the same material. Similar results were obtained for the assumed pavements in the moderate and cool climate. Slope variances increased as climatic temperatures increased. However, the increase was significantly less for the mixes containing sulfur than for the conventional asphaltic concrete.

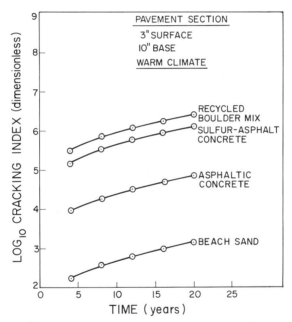

Figure 9. Comparison of log of cracking index for sulfur–asphalt and asphaltic concrete mixtures

The criterion for cracking of the pavement is based on fatigue resulting from the tensile strain at the bottom of the asphalt concrete or sulfur–asphalt concrete layer. The VESYS IIM program calculates a dimensionless cracking index. Although this index cannot be related directly to the area which is cracked, it can be used to indicate anticipated damage when comparing different pavement design sections. Figure 9 presents a graph of the logarithm of the cracking index as a function of age for the pavements with the 3-in. thick surface layers of all the trial materials in the warm climate. The trend of the data was similar for the other combinations of thicknesses, surfacing materials, and climates. The data indicate that the recycled Boulder mix and the sulfur–asphalt concrete are expected to suffer more fatigue cracking than the conventional asphaltic concrete mixture. However, the surface course made of beach sand mixture is expected to suffer significantly less cracking than all other mixtures even though this mix contains considerably more sulfur.

The overall structural adequacy of pavements is given in probabilistic terms of the present serviceability index developed from the AASHTO road test (17). This index, which is a measure of the momentary ability of a pavement to serve traffic, is based on such factors as rut depth, slope variance, cracking, and patching of the pavement. The relationship between serviceability index and these pavement distress modes is given by the AASHTO road test equation (16):

$$PSI = 5.03\text{-}1.91 \, Log_{10} \, (1 + SV) \, -0.01 \, \sqrt{C + P} \, -1.38 \, (RD)^2 \quad (11)$$

where PSI is present serviceability index, SV is slope variance, $C + P$ is the amount of cracking and patching on the paving surface, and RD is rut depth. This equation is programmed into the VESYS IIM program. Figure 10 presents the anticipated decline of present serviceability index

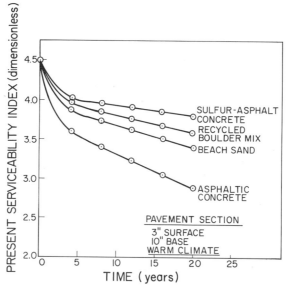

Figure 10. Comparison of present service ability index for sulfur–asphalt and asphaltic concrete

with time for pavements with the 3-in. thick surface layers in the warm climate. This figure indicates that the present serviceability index of the pavement surfaced with sulfur–asphalt concrete containing limestone aggregate is higher than those of the pavements surfaced with the other trial mixes. The pavement surfaced with the conventional asphaltic concrete mix had the lowest *PSI*. A similarity existed in the trend of the data for all other combinations of thicknesses, surfacing materials, and climates evaluated in this study. These data indicate that the pavements using surface mixtures containing sulfur are expected to perform better than those using conventional asphaltic concrete mixtures.

Performance Prediction Conclusions. The rutting potential is less for the pavement surface layers containing some form of sulfur–asphalt concrete than it is for conventional asphaltic concrete, whereas the potential for fatigue cracking is greater. The overall performance of the trial pavements depends upon the definition of performance. The present serviceability index, as determined by the AASHTO road test equation, predicts that the pavements with surface layers containing sulfur will outperform the conventional asphaltic concrete. However, for other definitions of serviceability where fatigue cracking of the surface layer is considered a more important factor in determining pavement performance, the conventional asphaltic concrete may prove superior.

A structural design analysis is only one portion of the entire pavement system design process for selecting an optimum pavement structure to serve a given set of loading and environmental conditions effectively. Construction and maintenance costs must be considered in the pavement system design process in a trade-off manner. Such trade-offs will be influenced by the availability and price differential between asphalt cement and sulfur and by the relative effect on performance of the pavement material properties. Although an economic evaluation was not within the scope of this paper, it appears that the current emphasis on research and development regarding use of sulfur in roadway pavements is justified. It appears likely that an outlet will be provided for the abundant sulfur, inferior aggregates may be utilized, and more serviceable pavements may result.

Sulfur Utilization in Recycled Asphalt Pavements

Recycling of old asphalt pavements is an energy-intensive activity which is being stimulated by two factors—the decreasing availability of total absence of quality aggregates in a number of regions around the country and the current price trends and projected demands for asphalt cement. These factors were discussed in the beginning of this chapter.

As noted previously, sulfur can function both as an aggregate and as an integral part of the binder. Other features which enhance its poten-

Table VI.   Recycled Asphaltic Concrete Laboratory Test Results

Sulfur (w/o)	New Asphalt (w/o)	Soften- ing Agent (w/o)	Resilient Modulus ($\times 10^6$ psi)	Bulk Specific Gravity	Air Voids (%)	Marshall Stability (lb)	Flow
			Boulder Highway Material				
0.0	0.0	0.0	4.867	2.288	6.8	11280	14
0.0	1.0	0.5	0.958	2.340	0.6	4870	21
1.25	0.25	0.0	1.807	2.330	2.8	8550	21
1.25	0.0	0.0	1.950	2.349	3.2	11570	16
			Nellis Boulevard Material				
0.0	0.0	0.0	4.846	2.294	6.8	9590	13
0.0	1.25	0.5	2.270	2.370	1.5	3800	23
1.25	1.0	0.0	3.584	2.384	1.2	7820	22
1.25	0.0	0.0	3.270	2.375	3.6	11770	14

tial in pavement recycling are its ability to lower the viscosity of heated sulfur–asphalt blends below that of the virgin asphalt (4, 5, 10) and to increase the stiffness of the recycled sulfur–asphalt aggregate mixture when it cools. Its ability to lower the viscosity is being explored as a possible means of eliminating costly softeners or plasticizing agents normally introduced to improve the workability of recycled asphalt pavement mixtures. New asphalt cement or softening agents are often required in the recycling process because the asphalt cement in the original asphaltic concrete pavement undergoes an age hardening process. The second feature of increased stiffness of compacted recycled asphalt pavements with sulfur is being inevstigated for city and urban streets where existing asphalt pavements made with emulsions and cut back asphalt cements tend to produce recycled mixtures with low stiffness. These pavements tend to rut and corrugate easily under the heavy braking, turning, and acceleration stresses characteristic of urban traffic.

The U.S. Bureau of Mines is currently sponsoring a study by Texas Transportation Institute to evaluate the use of sulfur in the recycling of existing asphalt pavements (14). For comparative purposes, laboratory specimens of recycled asphalt concrete pavement are being prepared by adding various combinations of sulfur, virgin asphalt cement, and softening agent. In this program, the sulfur and asphalt are added to the mixture directly as individual components and also in the form of a sulfur–asphalt emulsion. Laboratory tests are being performed to determine the physical properties of these mixtures. Typical laboratory test results for recycled mixtures used in this program are listed in Table VI.

Figure 11 is a graph of initial bending strain vs. number of cycles-to-failure resulting from flexure fatigue tests for two of the recycled mixes studied. For comparative purposes, typical graphs for conventional asphaltic concrete are also shown. Other characterization tests such as creep and repetitive load were performed also on specimens of recycled asphalt pavement as discussed previously.

Since field data are not available currently on the actual performance of recycled asphalt pavements incorporating sulfur, the laboratory test data were used in the VESYS IIM program to predict field performance under traffic. Results of this evaluation were presented in the previous section of this chapter. Figures 6–10 compared a sulfur-recycled mix with other sulfur–asphalt mixtures and a conventional asphaltic concrete. These results indicate a superior performance over asphaltic concrete in all performance categories except fatigue. Work is now in progress to improve the fatigue resistance of sulfur-recycled systems by using blends of sulfur with lower-viscosity asphalt.

Based on studies now in progress, it appears that the use of sulfur in recycled asphalt pavement produces significant benefits by increasing workability during mixing, reducing the amount of virgin asphalt and/or softening agent which must be added, and increasing stiffness after placement. By recycling old asphalt pavements, conservation of aggregates and asphalt cement is realized, and high-quality pavements can be constructed.

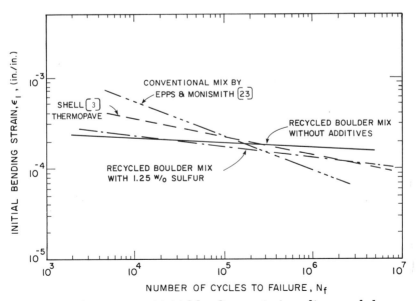

Figure 11. Comparison of initial bending strain for sulfur-recycled, conventional recycled, and conventional asphaltic concrete mixtures

Sulfur–Asphalt Pavement Field Trials

Various types of sulfur–asphalt pavement field test sections have been placed in the U.S., Canada, and Europe. Shell Canada Ltd. has conducted field trials in Canada (*1, 18, 19, 20*). Gulf Oil Canada has also placed test sections of sulfur–asphalt pavement in which the sulfur and asphalt cement were incorporated into the mixture as an emulsion (*4*). Société Nationale des Pétroles D'Aquitaine (SNPA) placed field test sections in France using the emulsion process (*5*).

In September 1975, a field test section incorporating various thicknesses and types of sulfur–asphalt concrete mixtures was constructed on U.S. Highway 69 near Lufkin, Tex. (*12*). Participating agencies were the Federal Highway Administration, Texas State Department of Highways and Public Transportation, Moore Brothers Construction Co., SNPA, Texas Transportation Institute, U.S. Bureau of Mines, Texas Gulf, Inc., Robertson Tank Lines, and the Sulphur Institute. The total length of the test section is 3600 ft. The aggregate–emulsion (AE) system was used to prepare the mixture. The binder was an emulsion consisting of 30 w/o sulfur and 70 w/o AC20 asphalt cement prepared in a portable colloid mill. The binder content ranged between 5.3 and 7.1 w/o of the total mix. The aggregate for one of the sulfur–asphalt concrete mixtures was a blend of 60 w/o pea gravel and 40 w/o local sand to comply with the gradation specification requirements of Texas Highway Department type D mix. Local dune sand was the aggregate used in another of the sulfur–asphalt concrete mixtures placed in the test section. This test section is now being monitored to evaluate the long-term performance under traffic.

The U.S. Bureau of Mines participated in a field trial of sulfur–asphalt concrete pavement on U.S. Highway 93 near Boulder City, Nev. in January 1977. This test section is 2100 ft long. The aggregate–asphalt–sulfur (AAS) system was used to mix the ingredients. The sulfur and AC 40 asphalt cement were introduced into the pugmill as individual components. The sulfur comprised 27 w/o of the total binder. The aggregate used in the mixture was a crushed volcanic rock which conformed to the Asphalt Institute type IVb gradation. This test section is now in post-construction evaluation.

Texas Transportation Institute participated in a demonstration field trial of sulfur–asphalt pavement in early April 1977 on U.S. Highway 77 in Kenedy County, Tex. This test section is 3000 ft long, and the mixture was prepared by the aggregate–asphalt–sulfur (AAS) system. The sulfur and AC 20 asphalt cement were introduced into the pugmill as separate ingredients. The total mixture consisted of about 79 w/o aggregate, 7 w/o asphalt cement, and 14 w/o sulfur. The aggregate consisted of a blend of about 35 w/o local dune sand and 65 w/o select coarse sand. This test section is also now in post-construction evaluation.

Literature Cited

1. "Sulfur: 1980, Shortage or Glut," *Chem. Eng.* (September 1976) 49–52.
2. McBee, W. E., Saylak, D., Sullivan, T. A., Garrett, R. W., "Sulphur as a Partial Replacement for Asphalt in Bituminous Pavements," "New Horizons in Construction Materials," pp. 345–362, Envo Publishing Co., Inc., 1976.
3. Deme, I., Hammon, R., McManus, D., "The Use of Sand-Asphalt Sulphur Mixes for Road Bases and Surface Applications," Proceedings Canadian Technical Asphalt Association, Vol. XVI, November 1971.
4. Kennepohl, G. J. A., Logan, A., Bean, D. C., "Conventional Paving Mixes with Sulphur-Asphalt Binder," Proceedings of the Association of Asphalt Paving Technologists, Phoenix, Ariz., pp. 485–518, February 1975.
5. Garrigues, C., Vincent, O., "Sulphur/Asphalt Binder for Road Construction," ADV. CHEM. SER. (1975) 140, 130–153.
6. Saylak, D., et al., "Beneficial Uses of Sulfur in Sulfur-Asphalt Pavements," ADV. CHEM. SER. (1975) 140, 102–129.
7. Sullivan, T. A., McBee, W. C., Rasmasson, K. L., "Studies of Sand Sulfur Asphalt Paving Materials," *U.S. Bur. Mines Rep. Invest.* (1975) 8087, 19.
8. U. S. Patent No. 3,783,853, June 12, 1973.
9. "Beneficial Uses of Sulfur-Asphalt Pavement," Volumes I-A, I-B, I-C, II and III, Final Reports on Texas A&M Research Foundation Project RF983, January 1974.
10. Texas A&M Research Foundation Contract No. RF3259, "The Extension and Partial Replacement of Asphalt Cement with Sulfur."
11. Dunning, R. L., Mendenhall, R. L., Tischer, K. K., "Recycling of Asphalt Concrete—Description of Process and Test Sections."
12. Saylak, D., et al., "Evaluation of Sulfur-Asphalt Emulsion Binder System for Road Building Purposes," Final Report on Texas A&M Research Foundation Project No. RF3146, January 1976.
13. "Cost Cutting Sulfur Pavement Making Inroads," *Eng. News Rec.* (September 30, 1976) 18.
14. Saylak, D., et al., "Recycling Old Asphalt Concrete Pavements," Proceedings of the Fifth Mineral Waste Symposium, Chicago, Ill., April 1976.
15. Predictive Design Procedure, VESYS Users Manual—An Interim Design Method in Flexible Pavement Using the VESYS Structural Subsystem, FHWA office of research, March 1976.
16. Rauhut, J. B., O'Quin, J. C., Hudson, W. R., "Sensitivity Analysis of FHWA Model VESYS II," Volumes 1 and 2, Report Nos. FHWA-RD-76-23 and FHWA-RD-76-24, March 1976.
17. "The AASHTO Road Test, Report 5, Pavement Research," Special Report 51E, Highway Research Board, Washington, D. C., 1962.
18. Deme, I., "Processing of Sand-Asphalt-Sulphur Mixes," Annual Meeting of the Association of Asphalt Paving Technologists, Wililamsburg, Va., February 1974.
19. Deme, I., "Basic Properties of Sand-Asphalt-Sulphur Mixes," the VII International Road Federation World Meeting, Munich, West Germany, October 1973.
20. Deme, I., "The Use of Sulphur in Asphalt Paving Mixes," Fourth Joint Chemical Engineering Conference, Vancouver, September 1973.

21. Fox, R. L., "Optimization Methods for Engineering Design," Addison-Wesley Publishing, Reading, Mass., June 1973.
22. Schmidt, R. J., "A Practical Method for Determining the Resilient Modulus of Asphalt Treated Mixes," *Highw. Res. Rec.* (1972) **404**.
23. Epps, J. A., Monismith, C. L., "Fatigue of Asphalt Concrete Mixtures—Summary of Existing Information," Fatigue of Compacted Bituminous Aggregate Mixtures, Special Technical Publication No. **508**, ASTM, 1972.

RECEIVED April 28, 1977.

Sulfur Composites as Protective Coatings and Construction Materials

J. E. PAULSON, M. SIMIC, and R. W. CAMPBELL

Chevron Research Co., Richmond, CA 94802

J. W. ANKERS

Chevron Chemical Co., San Francisco, CA 94105

Sulfur-based composites have been developed that have properties well suited for use as protective coatings and construction materials. Many of the elemental sulfur's useful properties are supplemented by additives that increase strength and durability. A variety of potential uses for sulfur composites have been demonstrated during three years of field testing, such as impervious liners for concrete ore leaching vats and for earthen irrigation canals and spill containment basins as well as stabilization of a slide-prone hillside. Spray is the preferred method for applying the molten sulfur composite although special equipment is required.

The unique properties of elemental sulfur make it a desirable base for coatings and construction materials. Among its attributes are hardness, resistance to chemical attack, high strength, and a low melt viscosity (*1*). Few, if any, common materials have this combination of useful properties. The commercial use of sulfur in these applications has been limited because of its brittleness, lack of resistance to thermal shock, and poor weatherability.

Technology to modify and control the crystalline nature of sulfur has advanced significantly in recent years. The beneficial effects of sulfur plasticizers, such as olefins and polysulfides, have been well documented (*2, 3, 4*). Fillers, reinforcing agents, and other modifiers can enhance further the desirable properties of sulfur (*4*). At the same time, they can improve its durability and thermal shock resistance.

The Chevron Chemical and Chevron Research Companies have conducted an extensive research and development program that has led

Table I.   Properties of a Typical Chevron Sulfur Composite

Property	Typical Value
Specific gravity	2.0
Viscosity at 285°F (cP)	600
Softening point (°F)	230
Coefficient of thermal expansion (10⁻⁶/°F)	25
Tensile strength (psi)	1200
Flexural strength (psi)	2100
Flexural modulus (10⁵ psi)	12
Compressive strength (psi)	7300

to a family of specialized sulfur composites called Chevron Sucoat coating compounds. Sucoat represents a new generation of sprayable sulfur composites with high mechanical strength and improved craze resistance. Equipment to make and apply them was also an outgrowth of the program. Key features of the compounds are rigidity, durability, good resistance to the action of weather and many corrosive chemicals, and ease of application. This chapter reports the status of Chevron's sulfur composites. It also describes test installations and application systems for some promising uses.

Composition and Properties

Chevron's emphasis has been directed mainly toward high-strength, sprayable composites which are chemically modified sulfur that contains organosulfur plasticizers, fillers, and other additives. The type and amount of each additive is selected to impart the properties required for each end use.

The properties of a typical Chevron sulfur composite are shown in Table I. One desirable feature is the low melt viscosity in the 250°–300°F range. This provides considerable latitude in the choice of application methods. Spray has proven to be effective for many uses. Brush, roller, and dipping have also been used for some coating applications. Most composites can be put into service immediately after application because they attain most of their strength and durability shortly after they solidify from the melt.

Sulfur's coefficient of thermal expansion can be reduced significantly when modified with plasticizers and fillers. Most Chevron composites have coefficients of expansion about one-half that of elemental sulfur. Even lower coefficients are possible with very high levels of filler. These lower coefficients contribute to the improved thermal shock resistance of sulfur composites and to their reduced tendency to crack or craze.

The mechanical properties of composites are good relative to other common construction materials. For example, they are 1.5–3 times as strong as a typical portland cement concrete (5). However, like all thermoplastic materials, their strength does vary with temperature. This requires that the relationship between strength and temperature be considered when selecting applications (Figure 1) (6). Composite strength reaches a maximum at about 40°–70°F. Unlike elemental sulfur, composite strength declines slowly as the temperature decreases. The modifiers present in the composite have some influence on the shape of the strength–temperature curve.

Laboratory and field testing has shown that properly formulated sulfur composites have good resistance to water, brines, and many acids (Figure 2). Acids and brines are mediums in which portland cement products often corrode severely.

Formulation has some influence on the resistance of composites to certain types of chemical solutions. In general, Chevron composites have about equal resistance to water, most mineral acids, and brines. In most cases, the composites have less resistance to oxidizing and basic solutions and to liquid hydrocarbons. However short-term contact can be tolerated usually without any serious detrimental effect on performance. In most circumstances, the composites have low resistance to caustic solutions, such as 5–10% NaOH.

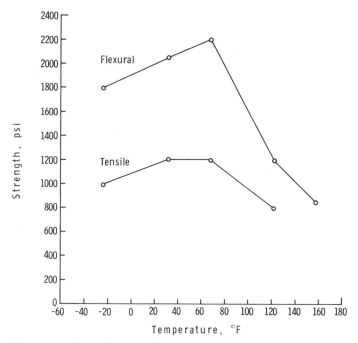

Figure 1. Typical strength of sulfur composites vs. temperature

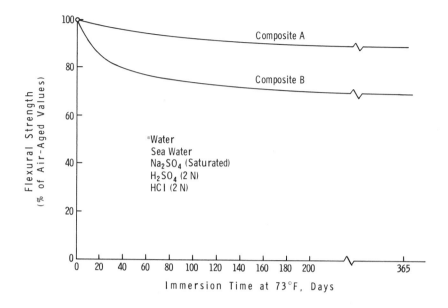

Figure 2. Chemical resistance of typical Chevron sulfur composites to water, sea water, Na₂SO₄ (saturated), H₂SO₄ (2N), and HCl (2N).

Equipment

New pilot and commercial-scale equipment has been developed to blend and apply the composites and to determine total system costs (Figure 3). Several commercial-sized units based on the pilot design are now under construction. Each has a 500-gal capacity. They are fitted with a hot-oil heating system, heavy-duty mixers, and heated hoses that can deliver up to 200 lb/min of sulfur composite. They are designed to operate typically at 250°–300°F and at spray pressures of 20–100 psi. Each is self-contained except for electrical power which is provided by a stationary source or a portable generator.

Portable batch plants have also been built to melt and blend composites on-site. These are useful for larger applications where multiple batches of composite are required. One such unit, built and maintained in Canada, consists of two steam-heated vessels mounted on a common skid. They can be used to convert sulfur into composites or to melt preblended composite delivered to the site in solid form. The vessels have capacities of 1500 and 800 gal. The pump and manifold system has a dual function: to transfer sulfur and composite to and from the vessels or to spray-apply composite directly from the tanks via heated hoses.

Uses and Applications

Part of Chevron's program has focused on the use of composites as liners for earthen structures. Their durability, strength, and ease of application offer important advantages over concrete and other lining materials for certain uses. Another phase of the program has emphasized the use of composites as protective coatings for concrete. In this application, their durability, weatherability, and corrosion resistance can extend the

Figure 3.   Equipment to blend and apply sulfur composites in the field. (top) Trailer-mounted, commercial spray unit used to prepare and apply up to 500 gal of composite per batch. (bottom) Skid-mounted batch plant has a capacity of 1500 gal of composite per batch. Composite can be transferred to mobile spray units or used directly from the mixing tanks.

service life of concrete in corrosive or severe weather service. The following examples illustrate these applications.

Canal Lining. Sulfur composite was used to line an earthen irrigation canal in southern Alberta, Canada (7, 8) (Figures 4 and 5). This installation was a joint project between Chevron, The Sulphur Development Institute of Canada, and the Eastern Irrigation District.

Two problems with the canal required correcting. One was excessive water seepage that made the surrounding farmland unusable. The other was heavy weed growth that reduced the water-carrying capacity of the canal and required that the canal be weeded periodically.

Sulfur composite was evaluated for this use because less site preparation was required than for concrete linings, the application equipment was mobile, and the composite could be applied potentially in the winter months when the canals were not in service. Concrete canal lining work is normally halted when the temperature drops below freezing.

The test site was the last 585 ft of a small distribution canal. Site preparation consisted of weeding and reshaping with a backhoe. A 4-in. gravel drain was placed on the bottom of the lower 340 ft of the canal to channel away any water that worked its way underneath the liner. The drain later proved to be unnecessary. Some hand trimming was required to smooth rough areas left by the backhoe and to dig small furrows on top of the berms. The furrows were used to key the edges of the liner to the soil to protect them from physical damage and to minimize water intrustion. Keying the liner later caused some cracking of the liner by

Figure 4. Sulfur composite used as a lining for an earthen irrigation canal. (left) Weed-choked canal before it was lined with sulfur composite. (right) Same canal in use with the sulfur composite liner.

Figure 5. Installation of the sulfur composite canal lining. (left) Molten composite was spray-applied to the earthen canal. (right) The canal after the composite lining was installed.

not allowing it to move independent of the earth during normal temperature cycling.

Access to the canal was reasonably good so the Canadian portable batch plant described earlier was used to make the composite and to spray-apply it to the canal. The batch plant, an electrical generator, and a steam boiler were placed on a 40-ft flatbed trailer. The trailer remained stationary while the composite was prepared and sprayed but was moved periodically between batches.

Several composite formulations were used at several thicknesses. They were applied to the canal in sections starting at the downstream end. The length of each section depended on the batch size and the lining thickness. Typically, about 225 ft were lined per batch. Design thickness ranged from 0.75 to 2 in. and was built up from a series of layers about 0.1 in. thick. Control joints were placed between batch sections and at various intervals within the typical 225-ft section.

Most of the composite was applied in 25°–45°F weather with no difficulty. Damp soil, ice, light coverings of snow, and shallow puddles of water presented no serious hindrance to attaining a continuous liner. Bubbles of vaporized moisture were trapped in the initial pass of molten composite, but subsequent layers went down smoothly.

After a full irrigation season the canal has remained free of weeds. Seepage has been reduced to the point where the adjacent land is usable. Some cracks in the lining have developed, but they have not effected seriously the performance of the liner. Analysis has shown most of the cracks to be the result of:

(1) Contraction stresses that developed during solidification

(2) Vertical displacement of the lining from frost heave

(3) Reduced load-bearing support caused by settling of the earthen base

(4) Weak zones in the lining where the thickness was below design specifications

Later test work has shown that the severity of cracking from these causes can be reduced significantly by changes in installation design and application technique.

Slope Stabilization. A rigid lining of sulfur composite was applied to the face of an unstable hillside (Figure 6). The hill had been cut to allow construction of several storage tanks. Exposed rock strata channeled rainwater to clay/shale sections below. These became slide-prone when saturated with water. Prior to installing the lining, several unsuccessful attempts were made to recontour the slope to improve the runoff patterns. However, just before the lining was installed, the first rains of the wet season caused another small slide on the face of the slope.

Figure 6. Slope stabilization with a sulfur composite lining. (top left) Slope became unstable when saturated from heavy rains. (top right) Spray application of sulfur composite to the rim of the slope. (bottom left) The slope after the lining was installed.

Sulfur composite was selected for this application over products based on portland cement and asphalt because of its weatherability, durability, and the minimal site preparation required. The lining was placed along the east and south rim of the hill extending down the slope about 130 ft, covering about 25,000 ft (2). Very little site preparation was done besides removing vegetation and large loose rocks. Anchoring devices and weep holes were not used, but they may be required for some applications.

A portable 1500-gal vessel was used to prepare the composite. It was positioned on top of the hill above the area to be lined. A pump and rigid manifold conveyed the molten composite 60 ft to the rim of the slope. From there a heated flexible spray hose conveyed it to the point of application.

A wooden platform was constructed to suspend two hosemen on the face of the slope. It was raised and lowered by winch-driven cables. An entire batch of composite could be spray-applied without stopping to reposition the winch and cables.

Composite was prepared from molten sulfur tankered to the top of the hill from a nearby loading rack facility. Pigments were used to match the composite with the surroundings since the slope is in a highly visible area. It was applied to the slope in sections that averaged about 1800 ft (2). Each section slightly overlapped the other, and no control joints were used. Lining thicknesses of 0.5–1.5 in. were evaluated. The fresh slide area was not coated, only the exposed strata above it.

The static head at the hose nozzle was sufficient to provide a 30- to 40-ft stream of molten composite at delivery rates up to 500 lb/min. The composite was rained down onto the slope. Care was taken to keep the nozzle moving in order to prevent the composite from running before it solidified.

The lining has been in place through most of two rainy seasons, and no more slides have occurred. As anticipated, some cracking has taken place because of the monolithic nature of the lining. However, the water shedding ability of the lining does not appear to have been affected adversely.

Slope stabilization is just one potential application for composites in the broad area of erosion control. Several other related applications are bridge abutments, freeway cuts, storage tank embankments, mine tailings, sand dunes, and areas of turbulent water flow.

Spill Containment Basin. Impervious liners for earthen basins is another use for sulfur composites. A spill containment basin surrounding several bulk storage tanks was lined as a joint project between Chevron, The Sulphur Development Institute of Canada, and The Environmental Protection Service of Canada (Figure 7). For this particular application, the lining requirements were durability, sufficient strength to support foot traffic, and resistance to an arctic climate.

Composite was applied to the berm and base surrounding a 35-ft diameter gasoline storage tank. The area totaled about 3800 ft (2). Lightly compacted fill was used to form the berm. No furrows were used to key the lining to the berm.

Figure 7. Spill containment basin lined with sulfur composite. (top) The basin surrounded a petroleum storage tank. (bottom) The sulfur composite liner was applied to the earthen berm and the basin bottom.

A skid-mounted batch plant complete with steam boiler and generator was placed on a flatbed trailer and was positioned adjacent to the site. Composite was prepared from slate sulfur and was sprayed directly onto the basin. Instantaneous spray rates were about 400 lb/min.

The lining was 0.75–1.5 in. thick and was built up from layers about 0.25-in. thick. Control joints were used to separate sections of different test design. They were also placed at locations of changing contour, such as along the base of the berm. The largest section was a quadrilateral of about 680 ft (2).

The installation was completed shortly before the onset of winter. A heavy snow covering has prevented us from observing the condition of the lining. However, it was essentially crack-free after its installation.

The techniques used to line the spill containment basin should also apply to other earthen structures. Examples include water catchment basins and sewage treatment ponds.

Concrete Coatings. The durability, weatherability, and chemical resistance of composites match up well with the required properties of protective coatings for concrete and other rigid materials. A typical application is the protection of concrete structures from the acid solutions used in metal ore leaching plants (9). In one such application, sulfur composite was applied to two leaching vats of a new experimental copper ore leaching facility (Figure 8).

The vats were constructed of precast concrete slabs. Each had a capacity of about 2600 ft (3). In service they were filled with crushed ore, pumped full of dilute sulfuric acid, held at ambient temperatures for about 10 days, drained, and cleaned of spent ore with a front-end loader. Unprotected concrete has a short service life in this use because of the corrosive action of the acid and the abrasive action of the ore and front loader.

Prior to application of the composite, the concrete surface was sand-blasted, washed with a 10% hydrochloric acid solution, rinsed with water, and dried. This ensured good adhesion to the fresh concrete surface. Spacers were placed directly over joints in the concrete and were removed after the coating was applied. The resulting joints were caulked with an acid-resistant sealant.

Composite thickness was a nominal 0.25 in. It was spray-applied at about 270°F with our 150-gal portable spray unit by building up to the design thickness with a series of thin layers.

The vats were put into service for 9 mo. During that time about 20 ore leaching cycles were completed. After another 12 mo of temperature cycling in the sun, no cracks or crazing were observed on the vat walls.

Other types of commercial coatings and membranes were also evaluated in the test facility at that time, and analysis indicated the performance of the sulfur composite was superior to all the others tested. At the conclusion of the test, all the other systems were in various stages of failure because of cracking, crazing, delamination, or tearing.

Two vats were left uncoated to serve as controls. They were severely abraded and corroded with much aggregate exposed. Up to 1 in. of thickness appeared to have been lost.

Figure 8. Sulfur composite as a protective coating for concrete ore leaching vats. (top left) Composite was spray-applied to the inner walls of the vats. (top right) In service, the vats were filled with ore and acidic leaching solution. (bottom left) View of a coated vat wall above the ore bed.

These results demonstrate the high level of protection that sulfur composites can provide to concrete and to other rigid materials. Examples of other possible applications are sewage treatment tanks and ponds, marine structures, pipe, and masonry block structures.

Summary

Chevron has demonstrated that it is commercially feasible to use sulfur composites as protective coatings and construction materials. This was done with large-scale field installations using special equipment and application procedures. Future efforts will focus on the commercial development of Chevron Sucoat products for these and other practical uses.

Literature Cited

1. Dale, J. M., Ludwig, A. C., "Mechanical Properties of Sulfur," "Elemental Sulfur," Chapter 8, The Sulphur Institute, Washington, D. C., 1965.
2. Barnes, M. D., "Aspects of Sulfur Research and Potential Applications," "Elemental Sulfur," Chapter 18, The Sulphur Institute, Washington, D.C., 1965.
3. Currell, B. R., Williams, A. J., Mooney, A. J., Nash, B. J., "Plasticization of Sulfur," Adv. Chem. Ser. (1975) 140, 1.
4. Dale, J. M., "Utilizing Sulfur-Based Spray Coatings," Min. Eng. (October 1973) 49–52.
5. "Concrete Manual," 8th ed., U.S. Bureau of Reclamation, 1975.
6. "Test for Flexural Strength of Plastics," ASTM D 790-71, "Test for Tensile Properties of Plastics, ASTM D 638 (modified).
7. Paulson, J. E., Ankers, J. W., unpublished data, 1976.
8. Underwood McLellan and Assoc. Ltd., unpublished data.
9. Pickering, I. G., Watson, J. A., Dale, J. M., Ludwig, A. C., "A Sprayable Sulfur Coating for Protection of Concrete Leaching Vats," American Institute of Chemical Engineers 78th National Meeting, Salt Lake City, Utah, 1974.

RECEIVED May 19, 1977.

Sulfur Foam and Commercial-Scale Field Application Equipment

R. W. CAMPBELL, G. L. WOO, E. P. ANTONIADES, and C. B. OLSON

Chevron Research Co., Richmond, CA 93802

J. W. ANKERS

Chevron Chemical Co., San Francisco, CA 94105

Newly developed sulfur foams were installed with pilot-scale field equipment in 1974 as road subbase insulation to prevent frost penetration at Anderson Road in Calgary and to prevent thawing of underlying permafrost at Dempster Highway near Inuvik. Surveillance of these installations shows satisfactory performance in both cases. A prototype commercial field unit which operates at temperatures as low as −50°F chill factor and lays foams at 800 lb/min has been built. The foaming equipment is housed inside a 43-ft long trailer, which is moved by a specially equipped tractor. The foaming mixture is discharged through multiple nozzles in the spreader at the rear of the trailer. Operation of this equipment was demonstrated recently in Calgary illustrating the practical application of sulfur foam in the field.

Sulfur foam can be used to protect the permafrost by using a combination of sulfur foam and local embankment materials for road, airfield, and other construction to reduce both the overall cost and the amount of embankment material that needs to be quarried in the arctic wilderness. The use of sulfur foam as subbase insulation prevents thawing of permafrost, which can cause subsidence during the warmer months. Another application uses sulfur foam with conventional pavement construction material for subbase insulation in frost-susceptible areas to prevent frost heave which is caused by freezing of the underlying soil.

0-8412-0391-1/78/33-165-227$05.00/0

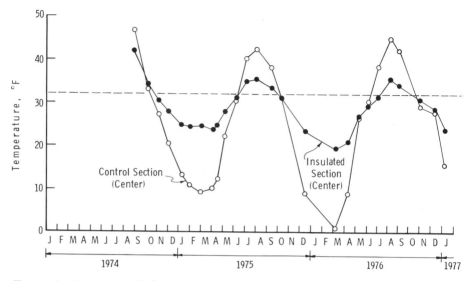

Figure 1. Dempster Highway test (near Inuvik). Temperatures below roadway
original ground level.

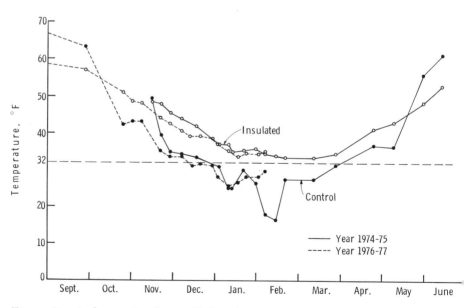

Figure 2. Anderson Road test (Calgary). Temperatures 3 ft below road surface a
center of control and insulated section.

Two field tests were conducted in 1974 to evaluate the above applications of sulfur foam. These tests were jointly sponsored by Chevron and the Sulphur Development Institute of Canada (SUDIC).

The first test, installed on a section of Dempster Highway about 40 miles south of Inuvik in the Northwest Territories, was designed for permafrost protection. The 45-ft wide and 130-ft long sulfur foam test pad averaged 4.4 in. in thickness and 11 lb/ft³ in density, with an average compressive strength of 46 psi. The foam was installed on top of the 1.5-ft thick pioneer fill and then was covered with 3.5 ft of granular overfill, in this case ripped shale, to complete the road section.

The second test, placed on a section in the east-bound lanes of Anderson Road in Calgary that had a history of frost heave problems, was designed to prevent frost heave. The sulfur foam pad, 140 ft long and 40 ft wide, was laid on the surface of the existing asphalt concrete road. The average thickness of the foam was 3.4 in. The average compressive strength of the 19-lb/ft³ foam pad was 163 psi. Following the application of a 1–2-in. leveling course of asphalt concrete, it was covered by 10 in. of asphalt concrete. The hot asphalt application caused no deterioration of the foam surface.

The sulfur foams for these tests were foamed-in-place with pilot field equipment at a foaming rate of 100 lb/min. Installation, instrumentation, and performance of these tests have been reported in detail (*1, 2*). Field surveillance data and sample analysis over two years so far indicate design insulation effectiveness and property integrity (Figures 1 and 2, Table I).

Table I.   Dempster Highway Installation Performance

	Original (1974)	Two Years in Service
Compressive strength (psi)	44 ± 5	41 ± 5
Moisture content (vol %)	0	0.5

Large-scale uses for this type of foamed-in-place insulation and application equipment are anticipated as the Canadian Arctic is developed. Successful results from these field tests and the prospects of commercial uses of the sulfur foam prompted Chevron Chemical Co. to proceed with the design, engineering, construction, and demonstration of a commercial-scale sulfur foam field application unit.

The Sulfur Foam System

The proprietary process for making sulfur foam was developed at Chevron Research Co. and uses carbon dioxide as the preferred blowing agent (3,4). Sulfur foams prepared by various methods and their applications have been described previously (5, 6, 7). Chemically modified sulfur is first made in the form of a concentrate (Chevron Furcoat concentrate), which can be mixed with elemental sulfur to make foam precursor as needed. Molten precursor is then mixed with a polyisocyanate foaming agent (Chevron Furcoat foaming agent) to produce foam. A very small amount of surfactant is also used.

Foam concentrate	+	Local molten sulfur	→	Foam precursor
(20–30 parts)		(70–80 parts)		(100 parts)
Foam precursor	+	Foaming agent	→	Sulfur foam
(85–95 parts)		(5–15 parts)		(100 parts)

Foam precursor can also be manufactured at a central plant. In this case, it is melted at the field site for use in the foam operation.

Sulfur Foam Properties

Sulfur foams with a wide spectrum of properties can be prepared by this process. Properties generally vary with density, which may range from 3 to 45 lb/ft^3. However, while density is kept constant, properties such as compressive strength, flexural strength, and closed-cell content may be altered by formulation changes. Some of the more common properties are listed in Table II.

The biological oxidation rate of sulfur foam depends on available surface area, temperature, and access of necessary nutrients, as well as on foam composition. Preliminary test data indicate that in all cases the oxidation rate is less than that of elemental sulfur. Soil pH around the foams at Dempster Highway measured two years after installation was never lower than the pH of the surrounding native soil (Figure 3). Sulfur foam's low toxicity is indicated by the LD_{50} of > 5 g/kg (rat) and the 100% survival rate in a fish bioassay with stickleback (96 hr) (8).

Commercial-Scale Field Application Equipment

A commercial-scale field foaming unit and melter–mixers have been built and demonstrated. Both were designed to operate at field temperatures as low as −50°F chill factor, which is considered about the coldest practical temperature for carrying out construction work in the arctic. Operation of the overall Furcoat system is illustrated in the Appendix.

Mobile Foaming Tractor and Trailer Unit. The components of the foaming unit are enclosed inside at 43-ft long trailer (Figure 4). The

Table II. Typical Properties of Rigid Sulfur Foams[a]

	ASTM Method	Density (lb/ft³)			
		4.5	*6.5*	*10*	*20*
Thermal condctivity at 86°F (Btu-in./hr-ft²-°F)[b]	D 2326	0.24	0.25	0.28	0.34
Compressive strength (psi)[c]	D 1621	25	40	50	170
Compressive modulus (psi)[d]	D 1621			~ 2,500	
Flexural strength (psi)	D 790			25	60
Flexural modulus (psi)[d]	D 790			~ 5,000	~ 5,000
Tensile strength (psi)	D 1623		11	9	24
Resilient modulus (psi)[e]				~ 16,000	~ 36,000
Coefficient of linear thermal expansion (°F⁻¹)	D 696			19×10^{-6}	13×10^{-6}
Dynamic loading (cycles to failure at 15 psi)[f]				$> 10^6$	
Water vapor permeability (perm-in)	C 355				
core		11	9	10	10
one skin intact		0.8	1.5	< 1	0.5
Water absorption (vol %)[g]	D 2127	1.5		2	2
Closed cell content (% of total voids)[a]	D 1940–62T	< 5	< 5	< 5	18

[a] Can be varied by chemical modifications.
[b] At −40°F, thermal conductivity for 10-lb/ft³ foam is 0.22 Btu-in./hr-ft²-°F.
[c] Measured parallel to foam rise, as maximum stress to 10% deformation.
[d] Very rough estimates.
[e] Compressive stress/strain under repeated loading conditions (0.1 sec loading at 20 applications/min).
[f] Compressions of 1.3 sec duration at 26 applications/min.
[g] Sensitive to hydrostatic pressure and closed cell content.

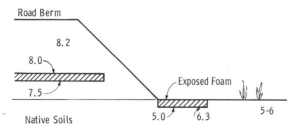

Figure 3. pH survey at Dempster Highway test installation (1976)

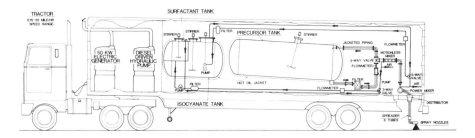

Figure 4.   Chevron Chemical Co. commercial sulfur foam unit

tractor is equipped with an auxiliary transmission which allows the
tractor to move the trailer at a ground speed of 0.15–0.50 miles/hr during
foaming. The unit can foam 21,000 lb of materials continuously before
refilling is required. Foaming rates range from 600 to 1100 lb/min.

Some of the major features of the Furcoat unit are:

(1) Jacketed vessels store the surfactant, polyisocyanate, and precursor.

(2) Electrically heated hot oil systems maintain temperatures.

(3) Valves and motors are hydraulically operated.

(4) Solenoids and a programmed timer provide automatic startup
and shutdown of the foaming operation.

(5) Flow indicators monitor the pumping rates of the three
streams during recycle and foaming.

(6) Flow rates and temperatures have digital readouts.

(7) A compressed air and storage system purges the foam components and cleans the mixer and downstream elements after foaming.

(8) A powerful ventilation–heating system and monitoring devices
assure clean air supply inside the unit.

(9) A spreader capable of discharging six streams lays a 10-ft wide
foam strip with a relatively flat surface.

(10) An intercom system allows communication by operators inside
the trailer, the driver in the cab, and the observer outside.

(11) A 50-kW electric generator supplies all the needed electric
power.

The three component streams are in a recycle mode at the desired
flow rates before foaming. During foaming, surfactant and precursor
are premixed before they come in contact with the polyisocyanate in
the power mixer. The foaming mixture is then discharged through
multiple nozzles in the spreader (Figure 5).

Portable Melter–Mixers for Foam Precursor Preparation. The
melter–mixers assembly (Figure 6) consists of a 1500-gal tank and a
800-gal tank having a capacity of 18,000 and 9,600 lb per batch, respectively. These jacketed tanks are equipped with stirrer, filters, circulating

pumps, valves, transfer hoses, and internal heating coils. The equipment is housed inside a portable winterized building.

Molten precursor can be prepared by melting either sulfur and foam concentrate or solid precursor. It takes about 2 hr to melt 18,000 lb of precursor in the 1500-gal vessel; this is enough for one full batch of foam. A 75-kW electric generator and a 50-hp steam boiler supply the power and heating needed.

Figure 5.   Sulfur foam being applied at 800 lb/min

Figure 6.   Skid-mounted melter–mixers having a combined capacity of 2300 gal being unloaded in the field

Demonstration of the Field Equipment in Calgary

On March 8, 1977 the Chevron Chemical Co. commercial field foaming equipment was demonstrated in Calgary to over 100 spectators representing local, provincial, and federal government, as well as local contractors and Canadian oil and gas companies (Figures 7, 8, and 9). This demonstration, jointly sponsored by Chevron and the Sulphur Development Institute of Canada (SUDIC), illustrated the practical application of sulfur foam in the field.

Two 10-ft by 150-ft strips of 6-in. thick, 7-lb/ft³ sulfur foam were foamed in place. The foaming rate was 700 lb/min, and the ground speed of the foam unit was 20 ft/min. A full batch (21,000 lb) of

Figure 7. Chevron Chemical Co.'s winterized commercial sulfur foam field application unit

Figure 8. Demonstration of Chevron Chemical Co.'s commercial sulfur foam field application unit in Calgary, March 8, 1977

Figure 9. Two 10-ft wide and 150-ft long strips of 7-lb/ft³ sulfur foam foamed in place side-by-side

7-lb/ft³ formulation is expected to lay a 10-ft by 600-ft, 6-in. thick foam strip. If the thickness of the foam were reduced to 3 in., the foam strip would be doubled to 1200 ft long. It was very windy during the demonstration. However, gusts of wind at 25–35 km/hr did not affect the foam application. The newly poured foam developed stress–relief cracks at right angles to the direction of laydown every 3–4 ft soon after cooling as observed in previous field tests. The surface of the foam strips was relatively even. The thickness of the foam can be varied and is determined by the foam formulation (density), the foaming rate, the width of the strip, and the ground speed of the foam unit.

Summary

Surveillance of sulfur foam test installation on the Dempster Highway in the arctic and at Anderson Road in Calgary continues, and in both cases the performance is satisfactory.

Commercial-size field equipment for the in-place foaming of sulfur foam suitable for operation at ambient temperature as low as −50°F chill factor was designed and constructed. It can be used to apply sulfur foam as subbase insulation for road, airfield, or building pad construction. Its operability was successfully demonstrated.

Appendix

Chevron Furcoat Arctic Insulation Systems Concepts

Barge Loading (Figure A-1). Chevron melter–mixers, special foam truck and trailer, plus all support equipment are loaded on a barge for shipment to the construction site. The major components, sulfur and foam concentrate, have been pre-mixed at the main plant to produce precursor to simplify remote site operations. The precursor, surfactant, and foaming agent, in bags and drums, can also be loaded at the same time. Enough materials are supplied to allow foaming throughout the entire winter season.

Melter–Mixer at Construction Site (Figure A-2). Precursor is loaded into the melter to be heated and melted.

Figure A-1.   Barge loading

Figure A-2.   Melter–mixer at construction site

Melting the Foam Precursor (Figure A-3). The melter–mixers and steam generators are set up inside a winterized building. Some of the raw materials must also be stored inside.

Loading Precursor into a Nurse Tank (Figure A-4). The hot foam precursor is loaded into a tank truck for delivery to the special foam unit. This step pays off when foaming is required at sites removed from the base of operations. The foamer is then used more efficiently with resulting higher lay-down rates per day. Drums of foaming agent are also delivered to the foam unit at the same time.

Figure A-3.   Melting the foam precursor

Figure A-4.   Loading precursor into a nurse tank

Transferring Materials to the Foam Truck (Figure A-5). Foam precursor, foaming agent, and surfactant are pumped into special tanks within the foam trailer with minimum down time. The foam unit is now ready to apply the next batch of Chevron Furcoat.

Applying Furcoat to Permafrost (Figure A-6). The Chevron foam unit is able to apply Furcoat insulation at rates up to 800 lb/min and at a ground speed of ¼–½ miles per hr. This unit applies the Furcoat in up to 10-ft widths.

Figure A-5. Transferring materials to the foam truck

Figure A-6. Applying Furcoat to permafrost

Backfilling with Gravel (Figure A-7). An overfill of gravel thickness enough to distribute the traffic load uniformly is placed directly on top of the Furcoat insulation and is leveled. The surface is now ready for use.

Other Chevron Furcoat Applications (Figure A-8). Chevron Furcoat may be used to insulate and protect the permafrost under air field landing strips and taxi ways, in pads for compressor stations, for insulation under pipelines and tank farms, and for a number of uses under and

Figure A-7.   Backfilling with gravel

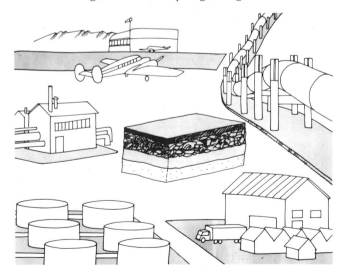

Figure A-8.   Other Chevron Furcoat applications

around construction sites, buildings, and housing. When the mobile foam unit is not being used for in-place foaming, it can be stationed in conjunction with a conveyor system to make board stock for later placement. This allows year-round production, eliminating the need to shut down during summer months or under adverse conditions.

Literature Cited

1. Campbell, R. W., Woo, G. L., Antoniades, E. P., Ankers, J. W., "Sulfur Foam: SUDIC-Chevron Field Test for Frost Heave Prevention," *Proc. Ann. Conf. Can. Tech. Asphalt Assoc., 20th* (1975) **20,** 195–210.
2. Campbell, R. W., Woo, G. L., Antoniades, E. P., Ankers, J. W., "Sulfur Foam for Permafrost Protection," "Materials Engineering in the Arctic," M. B. Ives, Ed., pp. 42–49, American Society for Metals, Metals Park, OH, 1977.
3. Woo, G. L., U.S. Patents **3,887,504** (June 3, 1975), **3,892,686** (July 1, 1975), and **3,954,685** (May 4, 1976).
4. Woo, G. L., Dale, J. M., Ludwig, A. C., U.S. Patent **4,011,179** (March 8, 1977).
5. Woo, G. L., Campbell, R. W., "Sulfur Foam—A New Rigid Insulation," Fourth Joint Chemical Conference of the American Institute of Chemical Engineers and the Canadian Society for Chemical Engineering, Vancouver, British Columbia, Canada, September 9–12, 1973.
6. Ludwig, A. C., Dale, J. M., "Cold Regions Applications for Sulfur Foam," Final Report, Contract No. DACA 89-71-C-0025, U.S. Army Cold Regions Research and Engineering Laboratory, February 1972 and U.S. Patent **3,337,355** (August 22, 1967).
7. Gamble, B. R., Gillott, J. E., Jordaan, I. J., Loov, R. E., Ward, M. A., "Civil Engineering Applications of Sulfur-Based Materials," *Adv. Chem. Ser.* (1975) **140,** 154–166.
8. "Standard Method for Examination of Water and Waste Water," 14th ed., American Public Health Association, Washington, D.C., 1976.

RECEIVED June 9, 1977.

14

Sulfur–Sand Treated Bamboo Rod for Reinforcing Structural Concrete

H. Y. FANG and H. C. MEHTA

Department of Civil Engineering, Lehigh University, Bethlehem, PA 18015

This paper presents the basic factors for selecting bamboo, the mechanism of bamboo–water–concrete interaction, and the sulfur–sand treatment of the bamboo used for reinforcement in structural concrete.

Bamboo is one of the fast-growing perennial grasses. It is found around the world in most tropical, subtropical, and even in temperate zones. There are approximately 550 different species of bamboo varying in size from 1 to 12 in. (2.54–30.48 cm) in diameter. The length can vary from 10 to 100 ft (3.04–30.48 m). Some bamboo can grow as much as 3 ft in a single day. In general, the tropical bamboo has a larger diameter and is taller; the cold-region bamboo is shorter and smaller in diameter.

Bamboo is considered a low-cost natural material and has been used in many parts of the world as a low-cost construction material for many years. The applications vary from housing to furniture, bridges, boats, decorative pieces, etc. Bamboo was studied scientifically for engineering purposes in 1914 by H. K. Chu at Massachusetts Institute of Technology (1). Similar studies were made later in China (2), Germany (2), France (3), Japan (4), the Philippines (5), India (6,7), Egypt (8), Colombia (9), and other areas (10,11,12). A detailed review of the historical development of bamboo engineering has been given by Fang (13).

This research showed that bamboo has high tensile, flexural, and straining capacities. The tests also indicated that the strength-to-weight ratios are excellent for tension, compression, and flexure in all types of bamboo (14). However, bamboo has three major weaknesses: low modulus, low bond stress, and high water absorption which leads to decay. These weaknesses are the major reasons why bamboo is not used widely in today's modern construction field. Many researchers have attempted to reduce the water absorption potential by using paint (2,6), asphalt emulsion (6, 10), coal tar (10, 12), cement (5, 6), and other techniques (8, 9, 10). Unfortunately, these proposed methods are ineffective, either because of cost or difficulty in application. Recently, a simple, low-cost method was developed (11, 15) using sulfur–sand treatment techniques which can increase the bond stress and modulus and can reduce the water absorption tendencies of bamboo culm for engineering uses. This chapter presents additional research findings on bamboo reinforcement in structural concrete. Focus will be placed on selecting the natural bamboo culm, on the mechanism of bamboo–water and bamboo–concrete interactions, on how the sulfur–sand treated bamboo rod can improve the natural weaknesses of bamboo, and on the sulfur–sand treated bamboo test procedures.

Basic Considerations of Bamboo for Use in Structural Concrete

In order to obtain the best results for bamboo use in structural concrete or in any engineering structures, the following factors should be considered:

(1) The age at which the natural bamboo culm should be selected to give the maximum strength—young (green) bamboo vs. seasoned bamboo

(2) The best time to harvest the bamboo in order to obtain the best performance, such as strength and natural moisture content

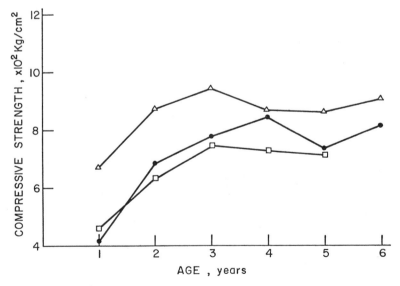

Figure 1. Effect of age on compressive strength of bamboo culm.
(●) Meng–tsung bamboo, (△) Makino bamboo, (□) thorny bamboo.

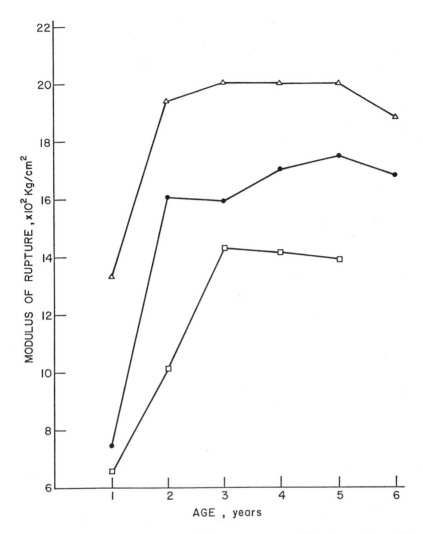

Figure 2.   Effect of age on modulus of rupture of bamboo culm. (●)
Meng–tsung bamboo, (△) Makino bamboo, (□) thorny bamboo.

(3) The mechanism of bamboo–water interaction and the swelling–shrinkage behavior of bamboo culm subjected to the dry–wet cycle

(4) The behavior of bamboo rod when placed in the fresh concrete and during the concrete curing process

Effect of Bamboo Age on Its Strength

Numerous factors contribute to the bamboo strength, such as type, size, age, storage conditions and duration, type of test and test procedures,

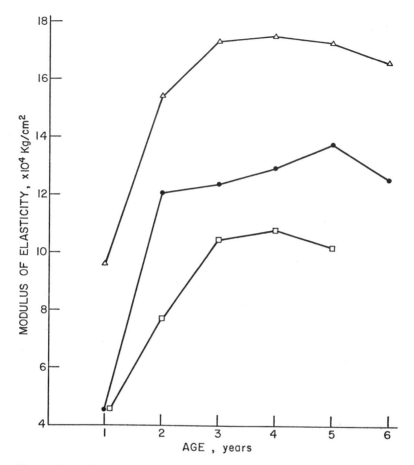

Figure 3.   Effect of age on modulus of elasticity of bamboo culm . (●)
Meng–tsung bamboo, (△) Makino bamboo, (□) thorny bamboo.

etc. Therefore, the results of bamboo strength tests are not consistent, and little published data is available. Recently, a comprehensive study on the effect of age, bamboo strength, and specific gravity was made by the Forest Department of National Taiwan University (16, 17). Typical results of compressive strength, modulus of rupture, and modulus of elasticity of bamboo culm vs. age are shown in Figures 1, 2, and 3. The results indicate that seasoned bamboo three to five years old will provide the highest strength. In addition, they concluded that the outer and the bottom parts have higher values for specific gravity than do the inner and top parts of the bamboo culm. The seasoned bamboo culm has a higher value than young (green) bamboo. This is also true for all three strength tests. A typical result for the modulus of rupture vs. location of bamboo specimen is shown in Figure 4. It was suggested that bamboo

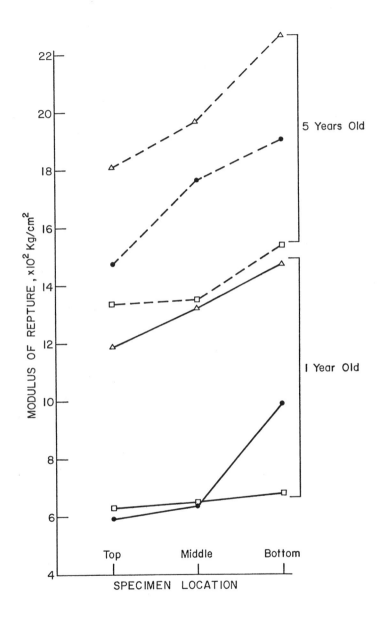

Figure 4. Effect of specimen location on modulus of rupture. Symbols are same as in Figures 1–3. Top = top portion of bamboo culm, middle = middle portion, bottom = bottom portion.

be harvested for use before serious biological and physical erosion takes place. The report includes the results of six types of Taiwanese local bamboo. Test specimen preparations and test procedures are described, and the test results are summarized in graphical form, together with statistical equations. These data are very useful guides for selecting the natural bamboo culm for engineering uses.

In order to evaluate the variation in the stress–strain characteristics for a given bamboo culm, a Pennsylvania bamboo was used for this study. All Pennsylvania bamboo was obtained in the vicinity of Lehigh University. The diameter of the bamboo culm varies from 0.5 to 2 in. (1.27–5.08 cm). The length is approximately 10 to 15 ft (3.04–4.56 m) and is classified as "A tecta" bamboo (10). Figure 5 shows the bamboo specimens used for the compression tests. Two strain gages were placed on the

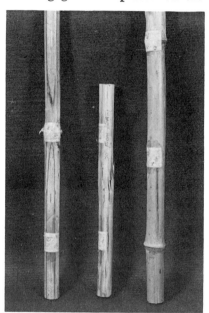

Figure 5. Test specimens of Pennsylvania bamboo. Photo also shows the locations of strain gages.

surface of each bamboo culm at different locations. Typical stress–strain relationships are shown in Figures 6 and 7. No significant differences between bamboo culm with knobs and without knobs were found.

Mechanism of Bamboo–Water Interaction

Bamboo readily absorbs water and swells. When dried it shrinks, consequently causing a loose bond at the bamboo–matrix interface which results in cracking. Therefore, the mechanism of bamboo–water interaction is very important because it will determine how bamboo will be

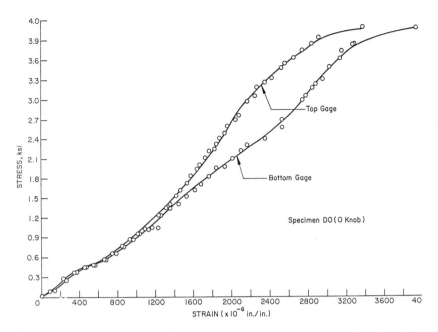

Figure 6. Typical stress–strain relationships of Pennsylvania bamboo (with no knob)

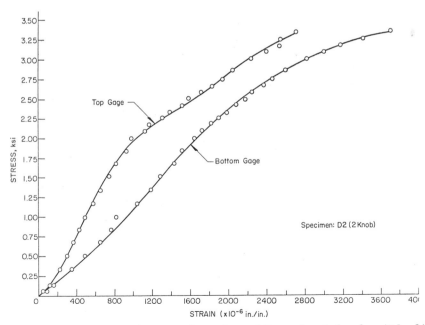

Figure 7. Typical stress–strain relationships of Pennsylvania bamboo (2-knob)

used in further engineering applications. To exchange the mechanism, the characteristics of water absorption and shrinkage of natural bamboo culm should be investigated. In general, the water absorption and shrinkage is directly related to the bamboo age, storage duration, harvest time, geological location, size, and type. The following sections discuss some of the findings on this behavior.

Swelling. Based on previous studies (2, 7, 8), bamboo culm can absorb up to 25% water in the first 24 hr. Figure 8 shows the percent of water absorption vs. soaking time in hours. Four types of bamboo were presented for the first 24 hr, and the percent of absorption varies from 20 to 40%. The general trend for all four types is similar.

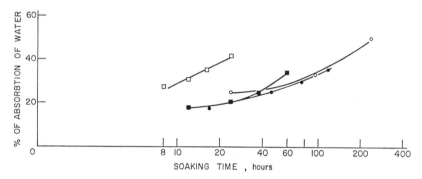

Figure 8.   Percent of water absorption vs. soaking time. (○) Japan (2), (■) Egypt (8), (□) India (7), (●) Pennsylvania.

Shrinkage. Bamboo has a large potential for swelling and shrinkage problems. There are three directions of shrinkage for a given piece of bamboo: radial, tangential, and volumetric. Test results from Japanese bamboo, cited by Wang (2), indicate that volumetric shrinkage is greater than that in the tangential and radial directions. The inner part of the bamboo shrinks more than the outer part of the bamboo. The tangential direction has very little shrinkage. Three types of Japanese bamboo were tested. All bamboo was air-dried for three months.

Harvest Period. Wang reported also (2) that bamboo should be harvested from late summer to mid-autumn because the natural moisture content of the bamboo is lowest then. Since the bamboo has not absorbed additional moisture from the ground, the swelling and shrinkage characteristics in the bamboo culm are at a minimum.

Mechanism of Bamboo–Concrete Interaction

Once the bamboo rod is placed in the fresh concrete, the bamboo absorbs the moisture from the concrete. The amount and rate of absorption depends on the bamboo properties and on the water–cement ratio. When bamboo absorbs the water, swelling occurs, and the volume of the bamboo increases. If the swelling pressure is large enough, it pushes the wet concrete aside. At the end of the curing period (approximately 21 days), the concrete hardens, and the bamboo loses the water and shrinks, leaving voids between the bamboo rod and concrete. The mechanism of bamboo–concrete interaction during the curing process is shown in Figure 9. The voids between bamboo–concrete trap air, moisture, and other foreign materials and accelerate the decay of the bamboo rod causing cracks in the structure. Mehra et al. (6) reported that some bamboo encased in concrete absorbs as much as 300% of its own weight in water and increases substantially in volume. Then it loses this moisture over a period of time and shrinks back to its original volume. Because of the swelling–shrinking behavior of bamboo rod, special treatment is necessary for engineering applications. The following section presents the basic principle of the sulfur–sand treatment method and procedures.

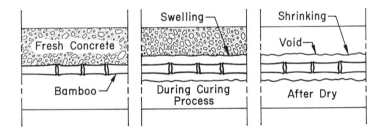

Figure 9.   *Mechanism of bamboo–concrete interaction during curing process*

Principle of Sulfur–Sand Treated Bamboo Rod

Sulfur is inexpensive, readily available, and has unique physical and chemical properties which make for a promising construction material as reported by Fike (18, 19) and others (11, 15, 20, 21). Rennie et al. (21) reported that the strength of sulfur varies from 200 to 1300 psi (1380–8970 kN/m²) depending on the temperature, as shown in Figure 10. Therefore, sulfur impregnation improves the natural weaknesses of the bamboo culm as used for engineering applications.

The main reasons for impregnating bamboo with molten sulfur are:

(1) To increase the confined pressure and reduce the cracks during loading conditions

(2) To waterproof the bamboo to minimize its swell–shrinking potential

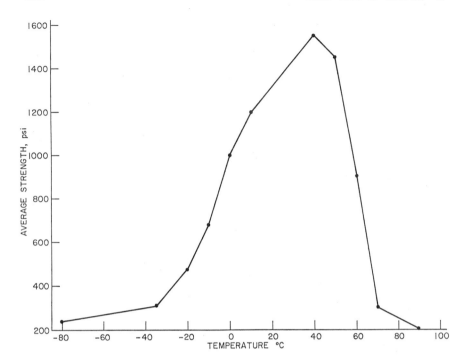

Figure 10.   Average strength of sulfur vs. temperature.   After Ref. 21.

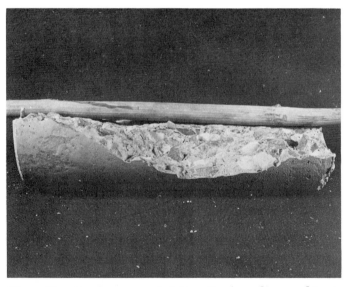

Figure 11.   Bamboo–concrete interaction (no sulfur–sand treatment)

(3) To aid the uniform coating of the sand on the bamboo surface

Before applying the sulfur to the bamboo surface, the bamboo's smooth skin must be removed by sand blasting; otherwise, the sulfur will not adhere to the surface. The thin wire wrapped around the bamboo culm, as shown in Figures 12 and 13, also helps to increase the confined pressure and to minimize the swelling potential.

Sulfur–Sand Treatment Procedures

All bamboo specimens used for the experiment were obtained from the vicinity of Lehigh University. Commercial sulfur in solid form was used. All specimens were air-dried for two months. All bamboo specimens were sand blasted, and thin wire was wrapped around the bamboo culm. The bamboo specimens were impregnated with molten sulfur at 280°–300°F (if the temperature is too high, it may burn the bamboo fiber) for approximately 1 hr, then air-dried. Before the sulfur-impregnated bamboo is completely dried, a coating of sand is applied to increase the bond stress and friction force between bamboo and concrete. The step-by-step treatment procedures are shown in Figure 14. Limited tests were performed on the effects of the number of impregnations on the bamboo rod (*see* Figure 13). It was found that the bond stress increases from one to two coats only slightly; however, increase in the flexural strength between one to two coats is significant. From the pull test,

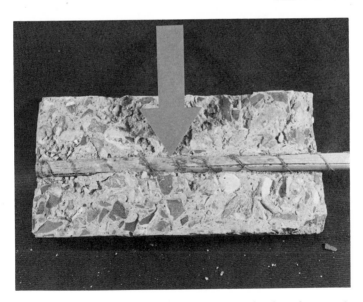

Figure 12. Bamboo–concrete interaction (with sulfur–sand treatment and wire)

Figure 13. Sulfur–sand treated Pennsylvania bamboo culms

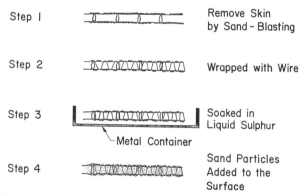

Step 1		Remove Skin by Sand - Blasting
Step 2		Wrapped with Wire
Step 3		Soaked in Liquid Sulphur
	Metal Container	
Step 4		Sand Particles Added to the Surface

Figure 14. Procedures for sulfur–sand treatment of bamboo rod

bamboo with and without sulfur–sand treatment indicate that the average value increases as shown in Table I.

Figure 11 shows that the nontreated bamboo has low bond stress and that the voids between bamboo and concrete are visible. However, for the sulfur–sand treated bamboo rod, there are no voids between the bamboo and concrete, and there is a high bond stress (15), as shown in Figure 12.

Table I. Comparison of Pull Test Performance by Treated and by Nontreated Bamboo Specimens

Bamboo Specimen	Pull Force (lbs)
Not treated	715–2868[a]
Treated with sulfur–sand (one coat)	2315–3100[a]

[a] Pull force range value depends on the number of knobs on the specimen and water–cement ratio.

Summary and Conclusions

(1) Bamboo between three to five years old has higher compressive strength, modulus of rupture, modulus of elasticity, and specific gravity.

(2) The lower portion of the bamboo culm has a higher strength value than the upper portion.

(3) The best time of the year to harvest bamboo is from late summer to mid-autumn, because at that time the natural moisture content of the bamboo, and therefore the swell–shrinking potential, is also low.

(4) Sulfur–sand treatment of bamboo is a low-cost, simple technique which gives bamboo higher strength, higher bond stress, low water absorption potential. Therefore, the bamboo can be used effectively for structural concrete and other engineering uses.

(5) Further study such as long-term performance of bamboo concrete, sulfur–bamboo–concrete interation, and effects of other additives in the sulfur are needed.

Acknowledgment

The work described in this chapter was conducted by the Geotechnical Engineering Division, Lehigh University as part of the research program on Development of Low-Cost Construction Materials. This particular study is sponsored by the Construction Materials Technology, Inc. International.

Literature Cited

1. Chu, H. K., "Bamboo for Reinforced Concrete," Thesis, Massachusetts Instiute of Technology, Cambridge, MA, 1914.

2. Wang, H. T. (Ed.), "Chinese Civil Engineering Handbook," Sections B-1 and B-5, 1944 (in Chinese).
3. De Simone, D., "Substitutes for Steel in Reinforced Concrete," *R. Eng. J.* (1940) **54**.
4. Shimada, H., "A Study of the Physical and Chemical Properties of Bamboo Poles," Transactions, Architectural Institute of Japan (1939–1940).
5. Purugganan, V. A. et al., "Research Study on the Use of Bamboo as Reinforcement in Portland Cement Concrete," Bureau of Public Highways, Manila, Philippines, 1957.
6. Mehra, S. R., Uppal, H. G., Chadda, L. R., "Some Preliminary Investigations in the Use of Bamboo for Reinforcing Concrete," *Indian Concr. J.* (January 1951).
7. Narayana, S. K., Abdul Rahman, P. M., "Bamboo Concrete Composite Construction," Journal, Inst. of Engineers, India (1962) **XLII**, (9) Part C15.
8. Youssef, M. A. R., "Bamboo as a Substitute for Steel Reinforcement in Structural Concrete," *New Horizons in Construction Materials* (1976) 1, Envo Publishing Co., Inc.
9. Hidalgo, O., *Bambu*, Estudios Technicos Colombianos Ltds, Colombia, S. A., 1974 (in Spanish).
10. Cox, F. B., Geymayer, H. G., "Expedient Reinforcement for Concrete for Use in Southeast Asia," U.S. Army Engineer Waterways Experiment Station, Technical Report C-69-3, 1969.
11. Fang, H. Y., Mehta, H. C., "Sulphur Impregnated Bamboo Pole," report prepared for Construction Materials Technology, Inc., Internat'l, 1975.
12. Glenn, H. E., "Bamboo Reinforcement in Portland Cement Concrete," Bulletin 4, Clemson Agricultural College, Clemson, S. C., 1950.
13. Fang, H. Y., "Bamboo Engineering," *New Horizons in Construction Materials* (1977) 2, Envo Publishing Co., Inc.
14. Fang, H. Y., Mehta, H. C., "Structural Response of Sulphur-Bamboo Reinforced Earth Mat to Seismic Loading," Proceedings, 6th World Conference on Earthquake Engineering, New Delhi, January 1977, Vol. 3.
15. Fang, H. Y., Mehta, H. C., Jolly, J. D., "Study of Sulphur-Sand Treated Bamboo Pole," *New Horizons in Construction Materials* (1976) 1, Envo Publishing Co., Inc.
16. Wu, S. C., "Studies on the Determination of the Cutting Rotation of Bamboo Based on Their Mechanical Properties," National Taiwan University in Cooperation with Taiwan Forest Bureau, Report No. 4, 1974 (in Chinese with English summary).
17. Wu, S. C., "The Effect of the Cutting Rotation of Bamboo on Its Mechanical Properties," *New Horizons in Construction Materials* (1976) 1, Envo Publishing Co., Inc.
18. Fike, H. L., "Sulphur Coatings, A Review and Status Report," Proceedings, Symposium New Uses for Sulphur and Pyrites, Madrid, Spain, May 1976.
19. Fike, H. L., "Surface Bond Construction," *New Horizons in Construction Materials* (1976) 1, 377–389, Envo Publishing Co., Inc.
20. Hubbard, S. J., "Feasibility Study of Masonry Systems Utilizing Surface-Bond Materials," U.S. Department of Army Technical Report No. 4-43, p. 20–22, 33–35, 1966.
21. Rennie, W. J., Andreassen, B., Dunay, D., Hyne, J. B., "The Effects of Temperature and Added Hydrogen Sulphide on the Strength of Elemental Sulphur," *Alberta Sulphur Res. Quart. Bull.* (1970) **7** (3), 47–60.

RECEIVED April 22, 1977.

Wood–Sulfur Composites

B. MEYER

Chemistry Department, University of Washington, Seattle, WA 98195

B. MULLIKEN

Molecular and Materials Research Division, Lawrence Berkeley Lab.,
University of California, Berkeley, CA 94720

Pure and modified sulfur can be used to impregnate, coat, and bond wood. Carefully prepared sulfur–wood joints have greater internal bond strength than whole wood. The mechanical properties of sulfur-bonded wood depend on the physical and chemical interactions at the wood–sulfur interface and on the chemical form of sulfur. Electron microscopy can be used to correlate the mechanical bulk properties with the microscopic nature of the bond interface, and laser Raman spectroscopy can be used to study chemical interactions. Good sulfur–wood bonds are obtained by coating wood with boiling-hot sulfur and quenching the product under light pressure. Basic research to control the chemical reaction between different wood components, sulfur, and possible additives is an important key in the development of this field.

Wood can absorb up to 100% of its dry weight in sulfur, yielding materials which are water resistant, durable, strong, and resistant to wood rot and fungi. This fact has been known for at least 60 years (1), but wood–sulfur compositions have not been studied extensively and have never been commercialized on any significant scale. This neglect is astonishing if one considers that both wood and sulfur are readily available and comparable in price. However, there are several reasons for this situation. Sulfur has mechanical properties which make it seem incompatible with wood: a density of 2.07 g/cm³ and a melting point of 119°C, a temperature at which wood loses moisture and resin. Below this temperature, sulfur is brittle while wood is elastic and resilient. Furthermore, sulfur and wood seem chemically incompatible because sulfur is hydrophobic and does not seem to wet or react with wood. Another factor is that sulfur chemists and forest products scientists work in traditionally far removed fields and face substantial barriers in vocabulary and training. However, the energy crisis, combined with environmental concerns, has created a unique situation which has increased interdisciplinary interest greatly.

The chemistry of the most important wood components has been studied for hundreds of years and is well reviewed (2). Wood adhesives and glues are also thoroughly studied. However, the chemistry of the presently leading commercial compositions still involves as much art as science (3). The status of the present knowledge of the properties and chemistry of elemental sulfur (4, 5) have been reviewed recently, and those of modified sulfur are reviewed in another chapter in this book.

The purpose of this chapter is to demonstrate the potential and limitations of research in the wood–sulfur field. It explores the interaction between the two materials and the properties of the resulting products. Earlier work on preparing sulfur–wood compositions is discussed, followed by descriptions of some of our work during the last 10 years. Finally, the chemical basis for wood–sulfur interaction is explored on the basis of the preceding knowledge. Other aspects of our work, including preparation of larger-scale specimens and application methods and techniques are published in two other papers.

Earlier Work on Sulfur–Wood Compositions

In the course of our work we discovered that wood pieces dipped in liquid sulfur form bonds which are stronger than those in whole wood. However, we were not the first to observe the beneficial effect of sulfur treatment. Kobbe (1) reviewed work performed before 1924. Unfortunately, neither he nor anyone else has listed references in this field; thus, we mention here some of the most important work involving wood or wood components. Ellis (7) used high-sulfur asphalt to introduce sulfur into wood. He also used hydrocarbons which can be saturated with up to 10% sulfur at 100°C to impregnate wood. Upon cooling, the solubility decreases, and the excess sulfur precipitates in situ. Weimar (8) pretreated wood with sodium carbonate and by dipped it into a hydrocarbon before soaking it in liquid sulfur and finally quenching it. Thus, he obtained an excellent impregnation. Kobbe (9) immersed wood for several hours into liquid sulfur at 140°–150°C until all water had boiled off. By regulating the immersion time he could control the weight gain. The final product was harder and tougher than the original material; the process was meant to upgrade soft and easily abraded woods, such as California Redwood. A possible use for this material was in railroad ties

(*10*) which were in increasing demand at the time. Hoskins (*11*) determined the absorption capacity of eight woods, including poplar, walnut, and hickory. Among the proposed uses for the wood were piano parts, engraving blocks, engineer's scales, acid tanks, buttons, billiard cues, bowling balls, and bowling pins; all these articles require high-density materials and smooth, tough surfaces. Ellis (*12*) proposed molding compositions comprised of wood, flour, starch, sulfur, and water for processing at 150°–200°C at "heavy high pressure." It was claimed that Cassava starch gave final products of unusual strength. This proposal and most of the later proposals reflect the earlier work quoted in Ref. 6. Darrin (*13*) used sulfur mixed with chlorinated diphenyls to impart hardness, rigidity, and resistance to wood products, including paper. Tillotson (*14*) used sulfur to waterproof wooden casting patterns. He used impregnation by immersion, by solvent penetration, and, finally, by chemical decomposition of sulfur containing compounds in situ. McKee (*15*) described a method for using sulfur to make strong corrugated cardboard. For this purpose, paper fiber was exposed to a sheet of liquid sulfur until its weight doubled. The resulting paper was corrugated while hot and processed with normal methods and adhesives. Jacquelin (*16*) used sulfur foam to make reinforced corrugated cardboard. Glab (*17–27*) proposed more than 10 different sets of compositions, all using a combination of lignocellulose and sulfur, together with a variety of additions. His goal was to produce high-quality products by pressing the shape at 200–1000 psi and up to 500°F. In some combinations a pressure of up to 10,000 psi was used (*18*). He developed, expanded, and perfected his composition over 15 years, but the basic concept and the final products matched those envisioned by Ellis (*28*). Several other compositions have been described; for example, a lignite product to be used as drilling mud (*29*) and a method for preparing polysulfide-impregnated wood chips for increasing the pulp yield of continuously digested lingnocellulosic materials. We cannot present a complete review of the field here and instead refer only to one more invention, that of Shapiro et al. (*30*), who used liquid sulfur with or without plasticizers, pigments, fillers, and flame retardants, as an adhesive to bond wood and animal skin. Neither this nor earlier papers or inventions describe the mechanical or chemical properties of the material in a systematic or quantitative way.

Experimental

It was the purpose of our work to provide a better definition of parameters and to identify the potential of the wood–sulfur system. For this purpose, we made a series of shapes, using a combination of wood sizes and materials, and reacted them with sulfur under a variety of conditions. In order to test the usefulness of wood–sulfur bonding properties, we decided to prepare samples by joining wooden parts with sulfur compositions so that we could test the mechanical properties of surface bonds at the interfaces and determine the weakest link between the materials by breaking the pieces apart and inspecting the failure surface. Two types of experiments were performed; a study of single bonds between two flat wood surfaces, similar to conditions used in plywood, and a study of molded samples containing a large number of small particles wetted with adhesive and compressed, similar to particle board.

Single Glue Line Between Flat Surfaces, Cut Parallel to Grain. A quick test to evaluate the tensile strength of the bond (*31*) is ASTM D-1037-72a. Samples prepared by bonding 5 × 5 cm wooden squares are mounted in an Instron instrument and are exposed slowly to tension until failure occurs. For many preliminary tests we chose five-ply commercial exterior grade plywood for cutting standard samples. Figure 1 shows a test specimen after failure. The tensile strength of our standard test sample, a commercial grade B-Ex five-layer plywood (*32*) sheet was between 31 and 63 psi, depending on the temperature, duration, and other parameters of treatment. Normally, specimens were conditioned for two days at 22°C and 65% relative humidity. If three consecutive samples failed in the commercial bond, whole pine wood blocks were bonded. The tensile strength of our samples of solid wood parallel to the grain was 100–300 psi.

PURE SULFUR. Elemental sulfur is a complex multi-component system. The molecular composition of pure liquid sulfur changes reversibly

Figure 1. Typical fracture of sulfur-bonded plywood sample, tested according to ASTM D-1037-72a. The failure occurred at 134 psi.

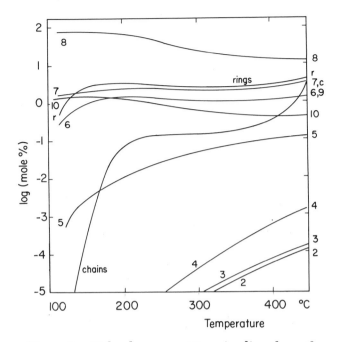

Figure 2. Molecular composition of sulfur above the melting point. Commercial press temperatures lie between 130° and 180°C. At 159°C the viscosity of sulfur changes, indicating polymer formation (4).

with temperature, Figure 2 (*4*). Thus, its wettability and other physical and chemical interactions change with temperature. Wood is an equally complicated system. However, its properties change irreversibly with temperature; above 100°C water vaporizes, and with increasing temperature wood degrades. The speed, mechanism, and product depend on temperature, exposure time, history, and the surrounding gases. Wood and sulfur in contact with each other undergo complex slow reactions which are sensitive to the type of wood used and to the history of the sample. For screening tests three specimens of plywood and full wood, usually sugar pine (*pinus lambertiana* Dougl.), were wetted or soaked in liquid sulfur at temperatures varying from 120° to 440°C, for 5 sec to 1 hr. After several hundred experiments involving soaking wood in sulfur for several days and using vacuum and other techniques at various temperatures and at different temperature gradients, we discovered that the most reproducible method for obtaining tensile strengths above 200 psi was to apply boiling-hot, liquid sulfur in a nitrogen atmosphere to wood which was preheated for 3–5 min at 130°C and to press the contact surfaces with approximately 10 psi. Seventy-five of 42 samples had a tensile strength of more than 100 psi perpendicular to the grain. Glue lines of up to 3 mm thick were tested and had comparable strength.

MECHANICAL PROPERTIES OF SAMPLES. The mechanical properties of wood samples are tested normally according to a series of ASTM tests (31) which are similar to those in Germany and France and are accepted world-wide. Plywood tests include tensile strength (D-897-72), shear by compression loading (D-905-45, 1970, revised), shear tension loading (D-906-64), and similar tests. Molded wood products are tested according to ASTM D-1307-72a, which includes various measures of internal bond strength, rupture, elasticity, and compression, as well as soaking and aging.

Originally, we tested only tensile strength. When we observed that it was comparatively easy to prepare glue lines stronger than solid wood, we discovered quickly that strong glue lines often are obtained at the expense of elasticity. As a matter of fact, we learned that soak tests in water weakened those samples which were stronger than others when dry. Practitioners in the forest product field recognize and try to avoid such failures, which are common in phenol–formaldehyde glue lines. This led to careful inspection of failure surfaces on samples glued with pure sulfur, modified sulfur, and commercial plywood samples, discussed below. An ideal glue line consists of a continuous transition between thin glue film and the whole wood with an intermediate layer into which the glue has penetrated to establish an elastic layer. This is necessary to strengthen the surface layer of the wood which is usually weakened by machining and to allow for adjustment to expansion and contraction during swelling.

ADDITIVES. In order to provide lasting, elastic glue lines, we tested a range of standard additives used in sulfur coating (3, 5) and in the forest product industry. Inspection of our glue lines indicated that the composition of the modified sulfur was not uniform. This could result from a variety of reasons. Careful analysis showed that wood resin is very soluble in liquid sulfur; thus, liquid sulfur extracts the resin, and upon heating a steady-state concentration gradient is obtained—steam and resin emerge from the wood, and sulfur and plasticizer penetrate the wood. Exposure of less than 1 min does not give sufficient sulfur penetration as emerging steam obstructs interaction; exposure of more than 3 min leads to supersaturation, with sulfur bleeding after quenching; and exposure of more than 7 min at 180°C or above leads to chemical reaction and embrittlement. Obviously, it is vital to control penetration carefully. This can be done by a judicious choice of time, temperature gradient, and concentration. A good, empirically determined compromise for small samples was given above. A better method would be to find a chemical process which interrupts the soaking of glue into the wood. In thermosetting glues this is achieved by proper formulation and pre-curing so that the viscosity temperature vs. time curve leads to gelling. This process control constitutes an essential art in the forest products industry. Despite 50 years of research and the sophisticated art, each batch must be supervised and tested individually. This is partly because of the diversity of woods but also because of the difficulties of premixing, aging, and storing. In order to exploit the experiences available in the conventional wood bonding field, we conducted experiments in which sulfur was introduced in situ into conventional phenol and urea resins used in the plywood and particle board industry (33).

Molded Samples. For this set of experiments a variety of commercial wood samples was used. The commercial fir or pine chips were graded with a ¼-in. screen before use. Bark shavings were washed, dried, and ground in some experiments; in others they were washed and used soaking wet. Normally samples of three sizes were prepared: round sticks, 12 mm diameter × 35 mm long; round sticks, 15 mm diam × 35 mm long; and flat samples, 5 × 5 × 0.5 to 1 cm. In some cases, round shapes with 3-cm diameter end tapered funnels were prepared to observe the influence of changing thickness. On standard samples the following parameters were tested.

SULFUR–WOOD RATIO. In each experiment a batch of approximately 100-g material was prepared by mixing ground sulfur and wood chips in such a ratio that the sulfur content constituted 0.2, 0.5, 1, 3, 10, 15, 30, 50, and 70 wt %. Since the density of sulfur is 2.07 g/cm^3, i.e., about three times that of wood, the volume and surface area of the wood was always substantially larger than that of sulfur. The moisture content of wood was normally about 10%. It was controlled by storing the wood at 22°C and 65% humidity. We found that attractive samples could be prepared over the entire sulfur–wood ratio range. However, the best sample preparation conditions were between 3 and 60% sulfur. Higher sulfur concentrations were difficult to handle and caused die corrosion. Under our crude laboratory conditions samples below 5% were overly sensitive to uneven mixing. This problem would not exist in large batches. Under commercial conditions sulfur contents of 5–10% might be ideal; in our small, 12-mm i.d. cylindrical dies, 50% sulfur samples were not reliably strong.

PRESSURE. Earlier work described molding pressures of up to 10,000 psi. Therefore, we built steel dies with 1-in. wall thickness. The strong walls provided a sufficiently large heat capacity to make it possible to press samples at 150°C without auxiliary heating with less than 10°C temperature drop during 3-min press time. Samples were packed into steel dies with round and square openings which were fitted with pistons on both sides. Standard laboratory pressure gages in the range of up to

15,000 psi were used. The pressure was manually applied for about 1 min or was slowly increased by motor driven plates which traveled at constant speed. Thus, the pressure increase was not linear. We tested a large range of pressures and found that the density of the product reaches a limiting value of about 1.7 g/cm³ at about 2000 psi, depending on temperature. In early parts of the work, high-density samples were desired to reproduce literature data; at a later stage low pressures were chosen, because it was found that high pressure is not necessary to form good bonds and because the present commercial interest is in low-density products with values between 0.6–0.8 g/cm³, comparable with whole wood which constitutes a compromise between insulating properties and mechanical properties at acceptable long-distance shipping costs.

We found that at high density and low sulfur concentration there is no decisive correlation between sulfur content and strength. As a matter of fact, we found that with many shapes and types of wood chips and under appropriate moisture and packing conditions no binder was necessary to form various-shaped articles.

TEMPERATURE. Four types of experiments were conducted:

(1) Sulfur and wood were mixed at room temperature and heated in the press.

(2) Wood was preheated in a dry box, and liquid sulfur was sprinkled over it before pressing into the cold die.

(3) Liquid sulfur was sprinkled over cold wood and cold pressed.

(4) Hot wood and cold sulfur, at room temperature, were mixed and pressed in a pre-heated die.

Without special equipment, it proved difficult to control the temperature of liquid sulfur during spraying; thus, methods (1) and (2) were preferred. Four temperature ranges could be distinguished. Below 110°C bonds were weak; between 110° and 150°C the bond strength depended strongly on time; between 150° and 250°C the bond strength went through a maximum; and above 250°C the sample changed quickly. The results are connected with the chemical changes which both wood and sulfur suffer in the above temperature region. This effect corresponds to the effect found in the glue line parameter discussed above and which is covered further below.

TIME. Since the bonding involves both physical effects and slow chemical reactions, the mechanical strength of samples depends strongly on the press time at high temperature. Press times of approximately 30 sec–10 min are best at high temperatures. Longer press times dehydrate wood and often yield brittle solids. Below 140°C press time of about 1 min yields full bond strength for both the liquid and solid sulfur range.

ADDITIVES. A large number of additives were tested, including several commercial and laboratory fillers, glues, and binders. Furthermore, several commercial stains, carbon black, paint pigment, and organic azo dyes as available in our lab were tested for compatibility and cosmetic effects. Several wood varnishes, resins, rosins, waxes, and glues were used to test their compatibility as die release and surface finishing agents and to see whether the unpleasant odor of sulfur-modifying agents could be suppressed.

Neutral Additives. In the course of our work, carbon black gave excellent, smooth surfaces. Samples containing 30% sulfur and 5% carbon black are unaltered after eight years. Originally, silicon grease, and later vacuum pump oil, were added as die-release agents. This was found to be unnecessary. However, as little as 1 wt % of such materials increases the tensile strength and water resistance measurably.

Aqueous Additives. In the forest products industry moisture content is used to facilitate heat transfer and control density. We found that a moisture content of 12% shortened the time necessary to obtain optimum bond strength. Among other liquids tested were 1% soap solution, glycerine, oil, and various commercial paper and wood glue mixtures. As little as 1% of materials which dissolve or react with liquid sulfur at the melting point could half the press time necessary to obtain maximum strength, without appreciably speeding up the degradation time. Thus, additives such as aromatics, hydrocarbons, alcohols, and ketones extended the useful press time range and made the sample quality more reliable.

Reactive Components. Many useful sulfur modifiers, such as phosphorus compounds, chlorides, and the like, singe wood and yield products which deteriorate after a period of between a few days to several months. Likewise, strong bases or acids attack wood. Thus, we tested only modifiers with a pH between 5 and 8.

Since it was our goal to prepare work for developing chemical adhesion, we also tested several sulfur compounds. Among these the thiokols are well known, both as sulfur modifiers and as wood sealants. Since thiokols do not wet wood, especially resinous wood, and since these compounds are commercially far too expensive for large-scale bonding applications, we made preliminary attempts to synthesize polymeric compounds with reactive groups. After an extensive search, during which pH-modifying substances were eliminated, we tested a variety of organic sulfur polymer families (4). We found that all of them were either excessively expensive, difficult to apply, or both, and most were not stronger on a per wt % basis than ordinary wood glues, unless their concentration was high enough to essentially provide a matrix for the wood. This situation might be quite different under commercial conditions, where a very even and thin glue spray can be applied. However, it occurred to us that a partial in situ synthesis of organic sulfur polymers might be feasible by premixing organic reagents with wood, sulfur, or both. For this purpose, we tested the reaction of aldehydes in aqueous solution or in carbon disulfide with different polysulfides in various proportions at 130°–200°C and with a press time of 1 to 15 min. The results of Raman spectra recorded on cured samples are described below. They indicate that both components react chemically and yield a suitable adhesive. This work (33) will be reported separately.

ODOR. Pure sulfur is free of odor; odor indicates impurities. While whole wood exudes a very pleasant odor, a large number of useful sulfur modifiers unfortunately exude an unpleasant odor. Often, modified sulfur contains some decomposition products which impart to the final material a faint but insiduous odor to which many people are quite sensitive. Likewise, commercial wood products exude an unpleasant odor from excess formaldehyde which remains unreacted and is released over several months.

Discussion

The data discussed above show that the tensile strength of sulfur–wood bonds exceeds that of whole wood. The durability of bonds depends on a combination of factors, especially elasticity. Some 12 standard modified sulfur samples are now eight years old and have retained their strength.

Figures 3–6 show typical failure surfaces. Figure 3 shows the surface of a sanded pine, parallel to the grain, magnified × 80. This and the fol-

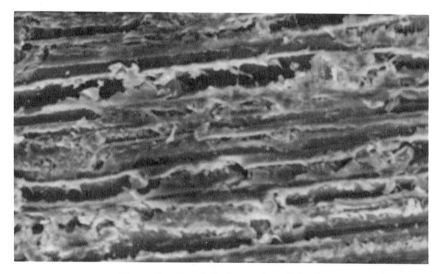

Figure 3.   Sanded pine surface (80 ×)

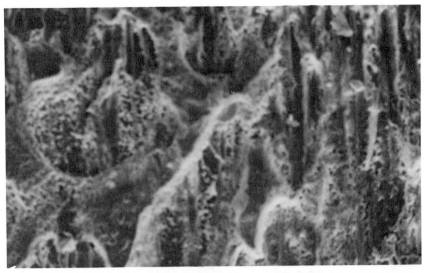

Figure 4.   Pine–sulfur failure surface (80 ✕)

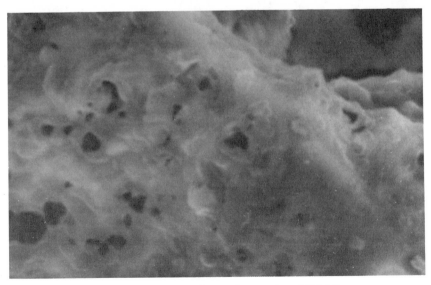

Figure 5.   Pine–sulfur failure surface (800 ✕)

lowing pictures were taken with a scanning electron microscope. The wood surfaces were vacuum shocked and gold shadowed according to standard sample preparation techniques. Figure 4 shows the failure surface of pine bonded with pure elemental sulfur (80 ✕). The lower parallel structure is the pine surface; the raised woven structure is the sulfur.

The failure occurred at 144 psi, and zones can be seen where an entire layer of wood is shaved. Figure 5 shows a section of the same piece in 800-fold enlargement. Figure 6 is an enlargement (240 ✕) of the fractured interface and shows that the wood surface has been disturbed. Some sulfur, in the upper right-hand corner, wets the wood. On the left hand side, a loose sulfur strand rests on a sulfur-wetted surface.

Figure 6.   Pine–sulfur failure surface (240 ✕)

The glue–wood interaction is complex. A good adhesive must have more than merely good inherent cohesive strength (*34*). One can distinguish four distinct points where cohesion is necessary. The intra-adhesive boundary, the wood–adhesive boundary where wetting and adsorption set in, the wood subsurface layer which has been weakened by tools or other mechanical or chemical treatment, and, finally, within the solid wood. If any of the five layers is rigid, the adjoining layer experiences extra stress and will likely fail upon loading. Since wood swells and shrinks with humidity changes, wood joints must be glued with an elastic adhesive which has the correct flow, transfer, penetration, and wetting characteristics to distribute itself over the entire interface area without excessive soaking. Current commercial wood resins tend to be brittle, and, especially with the phenol resins, failure of a "good bond" is regularly encountered at the wood subsurface layer. However, pure sulfur alone is so brittle that it is not suitable for bonding wood.

Hot liquid sulfur does react with wood components. Softwood contains about 40–45% cellulose, 20–30% hemicellulose, and 25–30% lignin, plus moisture and some resin, the chemical composition of which depends on the wood species. Hard wood contains about 5% more cellulose and 5% less lignin than softwood. We checked the literature and conducted exploratory experiments to investigate the basic chemical reactions between elemental sulfur and pure lignin and wood cellulose, and we found that chemical interaction is too slow to be significant in our systems. On the other hand, we found that it is almost impossible to find a wood sample which does not contain resin which bleeds into liquid sulfur. The most important components of wood extractives include: carbohydrates, organic acids, tannins, polyphenols, fats, hydrocarbons, resins, terpines, oils, and sterols. Of these, most compounds dissolve partly in liquid sulfur and form complicated multiphase systems. Above about 150°C all organic compounds react slowly with sulfur by a variety of mechanisms, involving rearrangements and bond scission. Above about 200°C almost all aliphatic substances dehydrate and decompose, forming hydrogen sulfide. Several of the wood extractives contain chemical functions which are relatives of known sulfur plasticizers. Thus, the exposure of wood to sulfur leads in situ to plasticization of sulfur. The concentration of plasticizer shows a gradient from the whole wood towards the bulk glue line. The nature of the gradient depends on temperature and press time. Thus, one can obtain excellent, lasting wood–glue bonds by reacting pure sulfur and pure wood.

Normally the wood surface contains a moisture film, and the cellulose surface is characterized by hydrogen bonding. Thus, it would be desirable to apply aqueous sulfur suspensions. The field of aqueous wood interaction has been well explored, chiefly in connection with wood pulping which involves many reactions which are similar to those feasible with elemental sulfur. Since it was not our goal to degrade wood, only neutral solutions were of interest, and occasionally a buffer was added to stabilize the pH. During a search for an economical chemical which would facilitate the introduction in an aqueous phase, we found that commercial wood glues, especially formaldehyde–urea and formaldehyde–phenol, are chemically compatible with sulfur under our press assembly conditions. Sulfur can be introduced into these compositions in different weight proportions and forms chemical compounds and intermediates which further react in situ during the wood processing, in the press, and which form strong, lasting, and moisture-resistant adhesives (*33*).

Figure 7 shows Raman spectra of samples of several reaction products. It shows that the sulfur-reinforced cured adhesive contains $-(CH_2-S)-_n$, polythiane links which have attractive properties. These compounds form under standard press conditions of 3–5 min at 200 psi at 150°C.

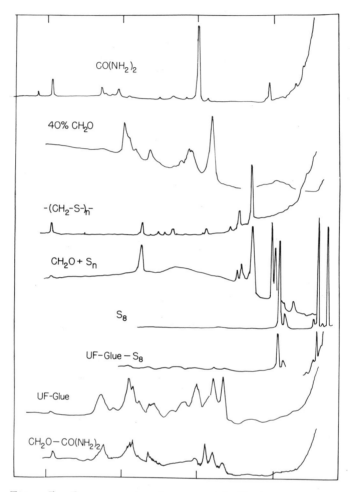

Figure 7. Raman spectra of commercial adhesive component samples. (top to bottom) (a) Commercial urea; (b) a commercial 40% formaldehyde solution; (c) polythiane, synthesized by reaction of sodium sulfide and formaldehyde (5); (d) the solid residue on a test sample bonded with formaldehyde–elemental sulfur in a 1:1 mixture; (e) pure elemental sulfur; (f) a commercial urea–formaldehyde wood glue reinforced with elemental sulfur; (g) commercial urea–formaldehyde wood glue without sulfur; and (h) a pure formaldehyde–urea mixture. All samples were cured at 150°C for 5 min.

Aging and Weather Resistance. Polymeric sulfur compounds are hydrophobic; thus, they have the potential to serve as exterior adhesives. However, in the case of wood, the lack of swelling can prove to be a handicap; in high-sulfur samples we observed poor water resistance and rapid crumbling. If the sulfur-to-wood ratio is small and if a large fraction of the wood chips remain unsealed, this can be overcome partly, and exceptionally good weathering conditions have been observed. The behavior depends on reproducible adhesive spreading which is achieved more easily on large scale than on our laboratory scale.

Commercial Potential. Currently, interest in the forest products field concentrates on finding a substitute weather-resistant adhesive to replace or complement phenol, which has undergone erratic price changes since the energy crisis in 1973. The practical criteria for new glues, in decreasing order of importance, seem to be: price; use with present production equipment and practice; a density of the assmbled product of about 0.7; a tensile strength of approximately 100 psi; and an acceptable color, odor, and touch. Furthermore, an ideal adhesive should require a minimum of mixing, a long assembly time (hopefully of several hours), a short press time, a low press temperature (if possible room temperature), and the ability to fill gaps up to 5 mm. The resulting bond should have an elasticity matching that of wood; strength to carry the required load; and durability toward heat, water, solvents, and sunlight.

Plasticized sulfur is a hot-melt adhesive and fulfills more of these requirements than any presently available, commercial glue. However, modified sulfur cannot be applied with present production equipment, and sulfur lacks many important, frequently required qualifications, among them fire resistance and resistance to heat. Thus, substantially more work would be necessary to make elemental sulfur a viable commercial bulk glue. However, we found that elemental sulfur can be polymerized in situ with formaldehyde resins (5, 33), and this yields bonds which have good mechanical properties, moisture resistance, and promising high-temperature behavior. Such glues can be handled with presently available equipment and presently common process conditions (35, 36). These materials are described separately (33).

This chapter was intended to awaken interest in a neglected field which contains great chemical challenges and potential applications. We have demonstrated that sulfur can be used to prepare bonds which are competitive in mechanical and chemical properties with those of the current commercial materials. If inherent problems with fire resistance and the threat of corrosion of manufacturing equipment as well as other problems of acceptance would be overcome, and if proper application machinery could be designed, sulfur might be useful as a viable bonding material. As long as the sulfur prices stay at the present level relative to other adhesive components, sulfur could well become competitive commercially in some sectors of the forest products field, especially those where large quantities of cheap adhesives are required. If the reaction mechanism and kinetics of the formation of thianes, phenylene sulfides, thiokoles, and other sulfur polymers was better known and if the properties of the compounds in pure form or in mixtures would be established under thermosetting conditions, an entire new family of high-quality, high-performance adhesives would become available which might warrant retooling of some plywood and particle board plants to use this glue in exterior applications.

Acknowledgment

During early stages of this work, A. Morelle, J. J. Smith, and S. Carlson participated in sample preparation. Much laboratory work was conducted by B. Paul-Gotthardt. Janice McOmber searched and collected valuable literature. Many samples were prepared at MMRD, Lawrence Berkeley Laboratory, U.C., Berkeley, where most testing was conducted. Some larger samples were pressed at the U.C. Forest Products Lab., Richmond, CA. W. Johns gave us many important scientific explanations; B. Bryant of the University of Washington College of Forestry participated in many valuable discussions and generously allowed us to use his hot press, his testing facilities, and other equipment in his forest products lab. Wood samples were donated by the American Forest Products Co., Martell, CA; bark shavings were given by Weyerhaeuser Co., Seattle, WA; and the sulfur was donated by Freeport Sulfur Co., Belle Chase, LA. The authors wish to thank Leo Brewer who made available to us the facilities of his materials lab.

Literature Cited

1. Kobbe, W. H., "New Uses for Sulfur in Industry," *Ind. Eng. Chem.* (1924) **16**, 1026.
2. Kollmann, F., Kuenzi, P., Stamm, C., "Principles of Wood Science," Springer, New York, 1975.
3. Bryant, B., "Wood Adhesives," in "Handbook of Adhesives," E. Skeist, Ed., Reinhold, New York, 1977.
4. Meyer, B., "Elemental Sulfur," *Chem. Rev.* (1976) **76**, 467.
5. Meyer, B., "Sulfur, Energy, and Environment," Elsevier, Amsterdam, 1977.
6. Currell, B. R., Williams, A. J., Mooney, A. J., Nash, B. J., "Plasticization of Sulfur," ADV. CHEM. SER. (1975) **140**, 1.
7. Ellis, C., U.S. **1,020,643** (1912).
8. Weimar, W., U.S. **1,375,125** (1921).
9. Kobbe, W. H., U.S. **1,599,135** (1926).
10. Kobbe, W. H., U.S. **1.599,136** (1926).
11. Hoskins, W., U.S. **1,631,728** (1927).
12. Ellis, C., U.S. **1,664,600** (1926).
13. Darrin, M., U.S. **1,889,088** (1932).
14. Tillotson, E. W., U.S. **1,927,076** (1933).
15. McKee, R. C., U.S. **2,568,349** (1951).
16. Jacquelin, G. J., U.S. **3,787,276** (1974).
17. Glab, T. W., U.S. **2,708,637** (1951).
18. Glab, T. W., U.S. **2,864,715** (1958).
19. Glab, T. W., U.S. **2,872,330** (1959).

20. Glab, T. W., U.S. **2,968,574** (1961).
21. Glab, T. W., U.S. **2,984,560** (1961).
22. Glab, T. W., U.S. **3,011,900** (1961).
23. Glab, T. W., U.S. **3,033,695** (1962).
24. Glab, T. W., U.S. **3,173,798** (1965).
25. Glab, T. W., U.S. **3,208,864** (1965).
26. Glab, T. W., U.S. **3,241,991** (1966).
27. Glab, T. W., U.S. **3,252,815** (1966).
28. Ellis, C., U.S. **1,665,186** (1928).
29. Kleppe, P. J., U.S. **3,874,991** (1975).
30. Shapiro, H., Giraitis, A. P., Sanders, R. N., U.S. **3,855,054** (1974).
31. *Book ASTM Stand.* **D-1037-72a.**
32. American Plywood Association, "U.S. Product Standard PS1-74 for Construction and Industrial Plywood," 1974.
33. Meyer, B., Johns, W. D., *Lawrence Livermore Lab. Rep.* **6913** (1977).
34. Mara, A. A., "Adhesive Properties," in "Treatise on Analytical Chemistry: Theory, Practice and Application," I. M. Kolthoff and P. J. Elving, Eds., Part 3, Vol. 2, J. Wiley, New York, 1977.
35. Fahrni, F., *Holz Roh- Werks.* (1957) **15,** 24.
36. *Ibid.* (1943) **6,** 277.

RECEIVED April 22, 1977.

INDEX

INDEX

The text of this book is set in 10 point Caledonia with two points of leading. The chapter numerals are set in 30 point Garamond; the chapter titles are set in 18 point Garamond Bold.

The book is printed offset on Text White Opaque 50-pound. The cover is Joanna Book Binding blue linen.

Jacket design by Diane Reich
Editing and production by Virginia deHaven Orr.

The book was composed by Service Composition Co., Baltimore, Md., printed and bound by The Maple Press Co., York, Pa.